EVALUATING THE EUROPEAN APPROACH TO RURAL DEVELOPMENT

Perspectives on Rural Policy and Planning

Series Editors:
Andrew Gilg, University of Exeter and University of Gloucestershire, UK
Henry Buller, University of Exeter, UK
Owen Furuseth, University of North Carolina, USA
Mark Lapping, University of South Maine, USA

Other titles in the series

European Integration and Rural Development
Actors, Institutions and Power
Michael Kull
ISBN 978 1 4094 6854 7

Rural Change in Australia
Population, Economy, Environment
Edited by Rae Dufty-Jones and John Connell
ISBN 978 1 4094 5204 1

Keeping it in the Family
International Perspectives on Succession and Retirement on Family Farms
Edited by Matt Lobley, John Baker and Ian Whitehead
ISBN 978 1 4094 0995 3

Rural Revival?
Place Marketing, Tree Change and Regional Migration in Australia
John Connell and Phil McManus
ISBN 978 1 4094 7511 2

Rural Housing, Exurbanization, and Amenity-Driven Development
Contrasting the 'Haves' and the 'Have Nots'
Edited by David Marcouiller, Mark Lapping and Owen Furuseth
ISBN 978 0 7546 7050 6

Evaluating the European Approach to Rural Development
Grass-roots Experiences of the LEADER Programme

Edited by

LEO GRANBERG
University of Helsinki, Finland

KJELL ANDERSSON
Åbo Akademi University, Finland

IMRE KOVÁCH
Hungarian Academy of Sciences and Debrecen University, Hungary

Routledge
Taylor & Francis Group
LONDON AND NEW YORK

First published 2015 by Ashgate Publishing

Published 2016 by Routledge
2 Park Square, Milton Park, Abingdon, Oxon OX14 4RN
711 Third Avenue, New York, NY 10017, USA

First issued in paperback 2018

Routledge is an imprint of the Taylor & Francis Group, an informa business

British Library Cataloguing in Publication Data
A catalogue record for this book is available from the British Library

The Library of Congress has cataloged the printed edition as follows:
Evaluating the European approach to rural development : grass-roots experiences of the LEADER programme / [edited] by Leo Granberg, Kjell Andersson and Imre Kovách.
pages cm. – (Perspectives on rural policy and planning)
Includes bibliographical references and index.
ISBN 978-1-4724-4376-2 (hardback)
1. LEADER (Program) 2. Rural development – European Union countries. I. Granberg, Leo. II. Andersson, Kjell (Professor of social sciences and rural research) III. Kovách, Imre.
HN380.Z9C6426 2015
307.1'412094–dc23

2014037366

ISBN 13: 978-1-138-54728-5 (pbk)
ISBN 13: 978-1-4724-4376-2 (hbk)

Contents

List of Figures

List of Tables

List of Contributors

Kjell Andersson is professor in Rural Studies at the Åbo Akademi University, Vaasa, Finland. He has been working with rural-urban and environmental issues for over two decades and has published extensively in three languages: English, Swedish and Finnish. His current research focuses on metropolitan ruralities. His books include *Beyond the Rural-Urban Divide* together with Erland Eklund, Minna Lehtola and Pekka Salmi (2009) and *Sustainability and Short-term Policies* together with Stefan Sjöblom, Terry Marsden and Sarah Skerratt (2012).

Almudena Buciega holds a PhD from the University of Valencia and is a sociologist at the University of Alicante, Spain. She also attained an MSc in rural and regional resources planning at the University of Aberdeen, Scotland. Between 1998 and 2006 she worked as main researcher in several EU projects at the Department of Geography (University of Valencia). Since 2011 she has been a lecturer at Florida Universitaria (Valencia). Her areas of interest include social dimensions of rural development, processes of social change and social capital, with particular attention to gender issues.

Aude-Annabelle Canesse, at Les Afriques dans le Monde, CNRS/Sciences Po Bordeaux, France, has been specialised in MENA countries since 2004, with a focus on Tunisia. Her activities combine academic activities (several fellowships) with practical experience in multilateral organisations. Based on public policy analysis, her work tackles development (institutions, tools, management), public administration and international scientific collaborations. More recently, she has analysed the dynamics and the potentials of conflicts in Tunisia after the Arab uprising.

Bernadett Csurgó is a PhD research fellow at the Institute of Sociology of the Centre for Social Sciences of the Hungarian Academy of Sciences. She is a sociologist and well experienced in rural sociology, elite studies and development policy. She has participated in several European research projects, with research focuses on rural-urban relationships, culture based rural development and gender issues.

Petra Derkzen holds a PhD in rural sociology from Wageningen University based on comparative research in rural governance and decision making processes. She was assistant professor at Wageningen University from 2009 until July 2013 in political sociology, food culture and food policy. Currently, she works as coordinator for Stichting Demeter, responsible for the Demeter biodynamic certification of farms and processed products in the Netherlands and Flanders.

Javier Esparcia is Professor of Geography and Local Development at the University of Valencia, Spain. He has been visiting scholar at University of Toronto, York University and University of Gloucestershire (CCHE). He has been advisor of the former Spanish Unit of LEADER Observatory. His main scientific interest includes change processes and socio-economic development policies in rural areas, such as LEADER. Currently, his research is mainly related to the analysis of social capital and territorial development, with particular attention to social networks, power elites and leadership in the dynamics of rural areas. Crisis and resilience in Spanish rural areas are also topics under research.

Jamie Escribano is an assistant lecturer in geography in the Department of Geography at the University of Valencia, Spain. He participates in several research projects on territorial development and social capital. Moreover, his research specialisation is on services in rural areas and their contribution to rural development. He had a couple of postdoctoral stays in French research centres in 2011 and 2013. He is member of the Local Development Institute of the University of Valencia, in which he develops management tasks in the Master Programme on Local Development.

Leo Granberg is scholar at the Finnish Centre of Excellence in Russian Studies, University of Helsinki; visiting researcher in Uppsala Centre for Russian and Eurasian Studies and professor for rural studies in social sciences at the University of Helsinki (2005-13). Currently his research focus is on social change in 'Second Russia'. He has studied rural development in former socialist countries, food systems, and rural-urban relations in late-modern countryside. He has edited the books *Sakha Ynaga – Cattle of the Yakuts* together with Juha Kantanen and Katriina Soini (2009), *Europe's Green Ring* together with Imre Kovách and Hilary Tovey (2001) and *Snowbelt, studies on the European North in transition* (1998).

Dénes Kiss is a lecturer in the Faculty of Sociology and Social Work at Babeş–Bolyai University of Cluj. He is also member of the Rural Studies Research Center from Cluj. His main research field is rural sociology, particularly the post-communist rural elite from Romania. He is also carrying out research activity in the field of sociology of religion.

Inna Kopoteva has been working as researcher in sociology at the Ruralia Institute, Mikkeli, University of Helsinki. She has conducted a number of studies in rural areas in Russia and in the Baltic countries. Her research projects include the development and implementation of administrative and municipal reform in the Russian rural areas, formation of partnership relations at the local level, and the development potential of local communities. Besides rural topics, she has studied entrepreneurship development in Russia and rural-urban migration.

Imre Kovách is a scientific advisor at the Institute of Sociology, Centre for Social Sciences, Hungarian Academy of Sciences, and professor of sociology at Sociology, Social Policy and Political Sciences Department, Debrecen University. He is honorary professor at Åbo Academy (Finland), visiting research fellow at the University of Helsinki and holder of the István Széchenyi Professorial Prize. He served as president in the European Society for Rural Sociology (2003–2007). His main research interests are integrated around the general theme of rural and agrarian sociology, and include political, social and economic elite, project-based economy and society, social integration, power relations, environment and climate-related issues. He is author and editor of over 200 books, articles and research reports.

Johanna Löyhkö worked as a project researcher on 'The Democratic Impact of Administrative Reforms – Temporary Government Instruments in Regional Development' in 2012. She also worked as a research assistant between 2009 and 2011 in two research projects: 'The Linguistic Consequences of the Solutions in the Public Administration' and 'LAGging behind or LEADER in Local Democracy? An Assessment of LEADER-type Development Projects as a Tool for Democratic Integration in the Contested Countryside'.

Ildikó Asztalos Morell is a sociologist working in the field of rural transitions in Hungary, with special focus on poverty, agrarian transformation and gender relations during state socialism and in the post-socialist context. She is currently associate professor at Mälardalen University and senior research fellow at the Uppsala Centre for Russian and Eurasian Studies. Her major publication is *Emancipation's Dead-End Roads?: Studies in the Formation and Development of the Hungarian Model for Agriculture and Gender 1956–1989* (1999). She has been co-editor for several international volumes including *Gender Transitions in Russia and Eastern Europe* (2005) and *Gender Regimes, Citizen Participation and Rural Restructuring* (2008).

Jouko Nikula works as a senior researcher at the Centre for Russian and Eastern European Studies at the University of Helsinki, Finland. He is also a scholar at the Centre of Excellence in Russian Studies. His research interests concern rural development, economic diversification and entrepreneurship in Russia and in the Baltic countries. He published his latest book, *Innovations and Entrepreneurs in Socialist and Post-Socialist Societies*, together with Ivan Tchalakov in 2013.

Marko Nousiainen is a political scientist working as a post-doctoral researcher at the University of Jyväskylä, Department of Social Sciences and Philosophy. Currently he works in a research project titled 'Participatory politics and democratic legitimation in the EU' (2013–15). He received his PhD in 2012, and in his doctoral thesis he studied the Finnish application of LEADER, concentrating especially on the aspects of governance, participation and political action.

Giorgio Osti is a rural sociologist and associate professor at the Department of Political and Social Sciences, University of Trieste, Italy. He is interested in the interplay between society and spatiality. He has been involved in research concerning local development in fragile areas and environmental issues like waste management and energy transition. He has published one article on the LEADER topic in the review *Sociologia Ruralis* (2000) and two papers in *Politiche, governance e innovazione per le aree rurali*, edited by Cavazzani, Gaudio and Sivini (2006), using the concepts of network and reciprocity.

Johan Munck af Rosenschöld is a PhD student in Environmental Policy at the Department of Social Research, University of Helsinki, and a researcher at the Swedish School of Social Science, University of Helsinki. His research is concerned with environmental governance, institutions and temporality focusing especially on projects as a means of organising. He is currently affiliated with the research project 'The Democratic Impact of Administrative Reforms – Temporary Government Instruments in Regional Development'.

Annette Aagaard Thuesen's current research lies in the empirical fields: The LEADER component of the EU Rural Development Programme and the Fisheries Programme, rural development in general and mobilisation and planning processes related to rural development. The theoretical approaches in her research are democratic network governance, partnership organising, meta-governance and institutional capacity. She has worked extensively on contract assignments and evaluation tasks for ministries, being in 2011 the project leader for the Danish Ministry of Food, Agriculture and Fisheries on the added value of the LEADER approach and studying in 2010 for the Ministry of the Interior and Health on the conditions for positive rural development. She has published articles in *Sociologia Ruralis, European Planning Studies* and *Local Government Studies.*

Enikő Veress graduated from the Faculty of History-Philosophy at the Babes-Bolyai University (BBU), Cluj, Romania. Her areas of scientific interests include social history, demography, urban/rural sociology, regional/local development, gender and community. She worked as a social researcher at the BBU between 1991 and 2012 and social expert at the Hungarian Unitarian Church between 2007 and 2009, performing and analysing researches in the field of regional/rural studies, historical demography and labour studies. Since 2000 she has participated in several international and EU research projects. She has published papers and articles in English, Hungarian, Romanian and German.

Chapter 1
Introduction: LEADER as an Experiment in Grass-Roots Democracy

Leo Granberg, Kjell Andersson and Imre Kovách

The Aim of the Book

The subject matter of this volume is the European rural development programme LEADER. The aim is to highlight this unique policy approach and to publish up-to-date research results on its achievements and limits, in order to discuss its merits and problems. LEADER is an initiative within the European Union's political repertoire which has been running for over 20 years. What makes it important is not only that it has a major role in rural development efforts, but also, that it has a pioneering role in the new type of governance that has been debated by policy-makers and political scientist over the last two to three decades.

Various questions connected to LEADER are taken up in the chapters of this volume, based on the experiences from different countries. At the local level, LEADER represents a new view on democracy, participatory democracy, compared with traditional representative democracy. Partnership sounds like an ideal way of working. However, many questions arise, not least the issue of power balance between unequal partners and the moral commitments they are willing to make, as remarked by Bernadett Csurgó and Imre Kovách (Chapter 4) in this volume. Even if LEADER has a large evaluation system and it has already been an object of some applied policy research (e.g. by the OECD), it is mostly monitored and studied in national circles and only in some cases in international comparison. An exception was a special issue of the journal *Sociologia Ruralis*, published in 2000 (Ray 2000).

Our aim is to take a look at the local level in the European Union and to study the way LEADER has responded to the challenges it was designed to address. This is not a mainstream evaluation report; our focus is not on analysing the monetary output of LEADER projects, or to calculate how the system has been used in different countries from the point of view of economic stimulation. Nor is our aim to evaluate the effects LEADER programmes and projects has on employment. Instead we are asking whether the LEADER approach strengthens local democracy or not, and how it affects the power balance among stakeholders, between national and local actors and between genders. Furthermore, we ask whether LEADER projects are indeed grass-root level activities, reflecting local needs and ideals, or if they are something else. We also consider how well the approach brings local know-how back to the development agenda, through innovations and development activities. Additionally, we examine if LEADER facilitates integrated local development and if its projects are connected to long-term beneficial development tracks or not. Finally we ask; how successfully is knowledge disseminated to other regions?

The Background of LEADER

LEADER is an acronym for the French: *Liaisons Entre Action de Developpement de l'Economie Rurale* (links between actions for the development of the rural economy) (European Commission 2006). The reason for the new rural policy tool was the concern in the European Union over the negative development of the countryside and the powerlessness and loss of perspective on agricultural policy in its attempts to solve the cumulating development problems in European rural areas. Agriculture had been a priority branch for the European Communities (later European Union) since its establishment. The lack of food after World War II, the strong political position of farmers and increasing prosperity made it possible to increase agricultural subsidies from decade to decade. The common agricultural market became the main political objective for this state union, not least when measured in the proportion of its budgets. The Common Agricultural Policy (CAP) was the policy implemented to reach this objective. The construction was shaken, however, by agricultural modernisation. Overproduction, increasing subsidies, outmigration from the countryside, as well as pressures from the changing global context, made changes in spatial policy unavoidable. In this situation, the LEADER approach was initiated in 1991.

Another development supporting the LEADER type of approach took place in the political sphere. Local action groups (LAGs) are the crucial agents in LEADER. They can be seen as local expressions of the shift from government to governance in European rural development policy, which is in-line with changes in many other policy areas with the objective of enhancing efficient and inclusive policy delivery at a local level. As Annette Aagaard Thuesen and Petra Derkzen (Chapter 8) argue in this volume, governance theorists have described the shift from government to governance as a change aimed to move decision-making increasingly onto multi-stakeholder platforms, and to decentralise central level decision-making to levels and arenas where knowledge and implementation resources are actually located. In contrast to 'government', the 'new governance' therefore indicates a pluricentric rather than a unicentric approach to governing, which also moves scientific analysis away from a state-centric approach (Rhodes 1996; Heffen et al. 2000). It is argued that governance implies an increased importance of networks as the principal means for social coordination (Sorensen and Torfing 2003), in which 'hierarchy or monocratic leadership is less important' (Van Kersbergen and Van Waarden 2004, 152). Moreover, the governance literature has put emphasis on processes that highlight the negotiation, accommodation, cooperation and formation of alliances that, after all, are the blood and bone of societal steering. All in all, many observers have evaluated the perceived shift towards decentralisation and broader participation in positive terms. Governance networks and 'partnerships' are seen as being capable of helping governments to deal more effectively with increased complexity and interdependency in society (Klijn et al. 1995; Rhodes 1996; Goodwin 1998; Bang 2003; Murdoch 2006). The new structures are said to improve the inclusiveness of decision-making by also integrating previously excluded groups (Shortall 2004, 113).

The structure of LEADER in the EU has changed from being a Community Initiative during LEADER I (1991–1993), LEADER II (1994–1999) and LEADER+ (2000–2006) to becoming a so-called mainstreamed element in the rural development programme (RDP) and fisheries programme (FP) of 2007–2013 (European Commission 2006). In the new member countries, some programmes similar to LEADER were applied before the EU accessions, for example the inter-municipal cooperation of the Regional Development Programme PHARE and the Special Accession Programme for Agriculture and Rural Development (SAPARD) in Hungary (Csurgó and Kovách, Chapter 4 in this volume).

EU support can be viewed as an expression of the need to introduce new players into the rural development scene. States cannot secure rural development alone, and the initiative has been transferred to other players, including actors from the private and the voluntary sectors. The most important actors in the LEADER programmes through the years have been the board members in Local action groups (LAGs). Accordingly, the core of the LEADER method is the establishment of the LAG partnership, consisting of representatives from the public, the private and the voluntary sector. These new players are important for the proper implementation of the decisions made by central authorities. They function as governance networks based on the idea that political power should grow out of empowerment that enable people to really participate, contrary to the limited participation that takes place through scheduled events such as formal elections (Bang 2005). According to the EU's basic guide publication, 'it was with the aim of improving the development potential of rural areas by drawing on local initiative and skills, promoting the acquisition of know-how on local integrated development, and disseminating this know-how to other rural areas' that LEADER was founded (European Commission 2006, 6). Taking the ideals built into LEADER as a given, it is time to ask whether these ideals have been realised and what sort of consequences this has had on the ground. Given the background and architecture of LEADER, organising, steering, democracy and power should in our view stand at the forefront of an examination of the programme. The ultimate objective of LEADER is rural development but rural development is insolvably intertwined with the issues above, which the EU, according to its statements, has been aware of. In the following we will present a more detailed theoretical and policy related framework for the examination of LEADER. We will begin with the concept of partnership, and then continue with local democracy and later on move to power structures. Following these, we will discuss the concept of projects, a central device and method in LEADER work, and finally we will take up the question of knowledge, one of the central concrete outcomes envisaged from LEADER work.

Partnerships

Partnerships between stakeholders are a most crucial part of the LEADER ideology, which is institutionalised in the structure of LAG groups. In contrast to traditional models of rural development, in which governments promoted their preferred developmental agenda through hierarchical bureaucracy or through market mechanisms, LEADER is aiming at building capacity among the local population with the goal of furthering common interests through common, coordinated efforts. The aim is 'to create public goods that will help to overcome the instances of market failure which characterise rural economies' (Kearney et al. 1994, 22).

Such a strategy is supported by many rural researchers, among others by Marsden (2008), who argues that the strength of networks within communities is intertwined with the potential of communities to grow. When communities are split along social, symbolic and cultural ruptures, they meet hindrances for growth and in getting out of the periphery. This is a topic developed by Denés Kiss and Enikö Veress in Chapter 10 on the difficulties of forming a LAG in Romania and by Ildikó Asztalos Morell in Chapter 11 on LEADER projects in the ethnically split Hungarian countryside.

When studying partnerships, it is worth analysing both the similarities and differences of interests, which local actors have.

In his classical work, Karl Polányi (1976) classified local actors into four types, which are the firm, the local state, civic society and the household, each of them having different basic interests. As Asztalos Morell (Chapter 11) notes, Söderbaum (2011, 49–50) contributes to the analysis of local collaboration by differentiating between competition-oriented and collaborative models. The first type is based on self-interest, with a focus on profit, and the survival of the unit often presupposes growth. The latter type is based on the principle of care for others and its aim is to achieve benefits for all the members within the network. Whereas the welfare states work along the constituency of a broad citizenship, civil society actions are not necessarily formed around universal interests. Söderbaum argues that a socially and ecologically sustainable society presupposes that collaborative models and ethical concerns are to be incorporated not only in the strategies of non-profit organisations, but even among profit-oriented companies. Nonetheless, even some idealistic organisation can be driven partly by commercial goals, even if reconciliation of commercial and collaborative interests will always be precarious.

For a successful partnership, the question of consensus or conflict is crucial. This question is much discussed in political theory and has a tight connection to the division made by March and Olsen (1989) into aggregative and integrative theories. As Marko Nousiainen (Chapter 6) argues, in aggregative theories, democracy means the aggregation or gathering of different views and preferences into a system of collective decision-making. The notion of conflict is emphasised in aggregative theories: democratic decision-making means choosing one of conflicting policies through political competition or bargaining. Integrative theories, on the other hand, emphasise deliberative action. According to the deliberative views of democracy the point is to form – through public discussion – such a collective decision that could be accepted by all. Rather than aggregating different preferences, deliberative democracy is about changing them. Habermasian ideas of communicative ethics and action especially are an important source of inspiration for theorists on deliberative democracy (March and Olsen 1989, 132). Experimenting with deliberative democracy may however turn into idealistic behaviour with quite surprising consequences. Csurgó and Kovách (Chapter 4) report that striving to avoid conflicts have led decision makers to organise meetings to negotiate planning, which as such sounds like good practise. However, to find consensus they have supported as many projects as possible, and in all settlements, to satisfy everybody, which leads to half-financed projects and many practical difficulties. According to Nousiainen's research (Chapter 6), in a Finnish LAG the urge to reach consensus was strong enough to curtail free and open discussion; instead it encouraged uniform thinking and even coercive means to achieve such thinking. In practise, entrance into the LAG was denied for those who did not share the dominating values.

Local Democracy Questioned

In Chapter 9, Giorgio Osti opens up a pivotal question, how to determine the will of the people (*volontà popolare*) through institutions considered to be models of democracy. This is crucial, because LEADER has often been marketed as a new, democratic way of local development. Results from our case studies are contradictory, indeed. For example Johan Munck af Rosenschöld and Johanna Löyhkö (Chapter 2) find that the LAGs in Finland and the UK are predominantly closed to external participation. They argue that this is problematic from a democratic point of view, as incorporating actors with less experience of project-based activities becomes less likely to succeed. In contrast, Csurgó and Kovách from Hungary as well as Javier Esparcia, Jaime Escribano and Almudena Buciega (Chapter 3) from Spain underline the importance of LEADER as a promoter of local democracy in their countries.

Osti addresses the question of whether the integration/aggregation dichotomy made by March and Olsen (1989) is useful for interpreting the issue, or if a triadic model would be more appropriate. His second question concerns local democracy: is it really broad enough a concept to include the material conditions (income, time, accessibility) of participation in the most important areas of public life? The question concerning breadth of participation relates to the debate on economic democracy – Italy fits well for researching these topics because of its strong tradition of municipal action.

March and Olsen's (1989) idea was to illustrate legitimated sets of rules useful for representing people's petitions. The 'aggregative pattern' is an institutional type of governance that resembles a market: numerous independent actors negotiate their different interests and achieve a substantial balance in the entire socio-political system. The political leadership acts as a sort of mediator among contrasting interests. The role of the public sector is therefore quite minor: it is required for control and for the distribution of very selective incentives. The 'integrative pattern' is another form of governance that recalls a community. The emphasis is on goods, values and destinies which are deemed to be in common and more important than individual interests. The crucial factor is a common cultural identity (Osti, Chapter 9). Thuesen (Chapter 5) – who analyses the rhetoric around the democratic capabilities of the LEADER system – argues that March and Olsen's (1989) two approaches can be viewed as an elaboration of the concepts of rational choice institutionalism and normative institutionalism (see also Bogason 2004, 3). March and Olsen themselves take the normative institutionalist stance:

> Political actors are driven by institutional duties and roles as well as, or instead of, by calculated self-interest; politics is organized around the construction and interpretation of meaning as well as, or instead of, the making of choices; routines, rules, and forms evolve through history-dependent processes. (March and Olsen 1989, 159).

The role of institutions and political leaders and the character of democracy differ in these two cases, and similarly the problems connected to each of the models differ. Thuesen opens these concepts further by listing three different types of integration and aggregation (Table 5.1, Chapter 5). Osti (Chapter 9) for his part goes further in order to show the defects in attempts to combine these two models and suggests that there is a need for something more, something he calls a third dimension to solve the dilemma of a well-functioning local democracy.

However, the concepts of aggregative and integrative democracy need not necessarily be viewed as normative models for how the popular will is formed and channelled into political decisions and societal steering; rather they may be used as analytical tools for studying quite universal aspects of democracy, the way Munck af Rosenschöld and Löyhkö do (Chapter 2). They combine the two conceptions of democracy with different aspects of the political process/analytical criteria, largely following the well-known input-output thought scheme: actors, institutional linkage, form of participation, conflict resolution, types of knowledge, outcome and accountability (Table 2.1, Chapter 2). This sort of analysis can be pursued without denying problems that none of the models can grasp, nor solve. The analytical problem is merely that general real-life situations tend to be a mix of aggregative and integrative political behaviour and that the research questions therefore should be very specific in order to sort out where the respective behaviours hold sway and what their merits and problems are.

Power Structure

The dilemma emphasised by Osti leads to a focus on local power structures. The LEADER approach changes the structure of local actors and their mutual power relations. The old dominant network of farmers and their organisations is on the retreat and a new 'project-class' of local developers is strengthening its position. As Kovách and Kucerova (2006 and 2009; see Chapter 4 in this volume) have argued, different experts, designers, European and national administrative staff, holders of intellectual capital as well as representatives of civil society occupy new social and class positions.

Being an instrument to foster local democracy, it is evident that LEADER has an initial democratic deficit, not least because a proportion of the LAG members, involved in decision-making bodies, are non-elected. In the same way, networks of governance, such as those derived from LEADER, are sometimes seen as undemocratic, due to the delegation of decision-making power to public, private and civic stakeholders (Thuesen 2010).

The bottom-up approach of LEADER is heavily emphasised in the literature. In its practical implementation, however, LEADER also has an important top-down component because of the strong role of the government in funding, planning and setting the rules for national LEADERs. Many of the practical inconsistencies in the application of LEADER are precisely due to this combination of top-down and bottom-up approaches. However, these inconsistencies may decrease with the increase of the complementarities between the two approaches, as Esparcia, Escribano and Buciega remark in Chapter 3.

Nevertheless, probably the most relevant interpretation of the 'negative externalities' of LEADER is connected to an analysis of power as a matter of social production, in the context of new rural governance. From this point of view, LEADER could be interpreted as the scene in which actors and institutions attempt to gain capacity to act, by blending their resources, skills and purposes into a viable and sustainable partnership. Sometimes this intended partnership is enmeshed with a paternalistic tradition which may explain the uneven distribution of stakeholders in LEADER and its decision-making bodies (Goodwin 1998). Frequently, however, new governance mechanisms deliberately seem to be used for the purpose of ensuring the continued hegemony of (some) local elites (Kovách 2000; Kóvach and Kucerova 2006). This objective may imply a tendency to involve (especially in the decision-making bodies) only the key actors belonging to or coming from specific elite groups (public, economic or civic, or a combination of these).

In their three case studies Csurgó and Kovách (Chapter 4) analyse who the actors are in the project class of contemporary Hungary. In this endeavour they also support Esparcia et al.'s (Chapter 3) remark that LEADER is not really following bottom-up principles. In Hungary, the control by the national bureaucracy is tight and the possibility of making a profit out of projects may sometimes tempt professional non-local actors to get involved and devalues the bottom-up character of the approach.

Gender divisions in the LEADER activities in Denmark are studied by Thuesen and Derkzen (Chapter 8). A European database from 2004 shows a clear male domination in the LAGs. In eight countries the majority of the LAGs had less than 25 per cent women on their boards. In six countries, the bulk of the LAGs had 25 to 50 per cent women on their boards. Only two countries had a female majority in most of their LAG boards. There are several possible reasons why fewer women than men are active in the LAG boards. Bock and Derkzen (2008) have outlined four different barriers to women's participation in rural policy making: (1) women's position in rural society and their weak socio-economic and political integration, (2) a traditional gender ideology that underlines women's domestic responsibilities and civil and apolitical involvement in the community, (3) the dominance of agriculture and economy in the rural development discourse and (4) the lack of fundamental structural and cultural changes in new governance arrangements. All these enumerated reasons add to the disadvantaged position of women in the public sphere. And they also give a clue regarding the barriers faced by other disadvantaged groups in getting a foothold on the new local governance system – discussed among others by Asztalos Morell in her review of the ethnic division in Chapter 11.

LEADER promises a shift in the power structure in rural policy from the national context to the local level. Such a move is also much needed in countries outside of EU. Aude-Annabelle Canesse (Chapter 12) describes and discusses the place of the local level in Tunisian development policy. Contradictory practises and weak results demonstrate the need for a basic rethink. Leo Granberg, Jouko Nikula and Inna Kopoteva (Chapter 7) analyse the results from an experiment with LEADER in Russian Karelia 2011–2013. It turned out to be possible to stimulate local action and to construct partnerships at the local level. This result was reinforced during field research in September 2014, in spite of ongoing changes in political situation in Russia. The question remains whether the regional and federal level authorities would be interested in the long run in backing such an activity. The possibilities of LEADER at the local level are underlined by Esparcia et al. (Chapter 3) who maintain that LEADER in Spain meant a real change of mentality in disadvantaged rural areas and managed to be a genuine tool for development. This result is based first on the territorial approach and second on the practical tools which LEADER offers. At the same time the writers are critical regarding the democratic deficits in LEADER, because some LAG members are non-elected, and because the combinations of bottom-up and top-down elements cause practical inconsistencies in the system.

Project LEADER

This new rural policy approach has delegated planning rights and generally the initiative to the local level, and organised the whole activity into a project-based political system, facilitating small local projects, sub-regional development programmes, and partnerships between different types of local stakeholders. Such a 'project policy' is one of the main alternatives in the new governance models, along with such organisational forms as 'public-private hybrid organisations', 'networks', and 'task forces'. They all fulfil functions that earlier were confined to classical public organisations or their counterparts in the corporate or civic sectors (Andersson and Kovách 2010, 7).

In this volume, several authors continue the research stream on projects, which started around 2003 in Helsinki in the Swedish School of Social Science at the University of Helsinki and continued in the SUSTAIN network (Sjöblom et al. 2006; Rantala and Sulkunen 2006; Kovách and Kucerová 2006). Earlier, project-based activity was used more in research than development and more in developing countries than in European countries. State funding especially did not take the form of projects with the exception of funding research projects. It seems, however, that the European Union had to find a new way to channel its activities and funding, and at the same time to avoid the transfer of its funding directly to member states' budgets, as well as to avoid competing directly with the funding streams from national budgets. Projects gave an opportunity to solve these two issues and to differentiate between European funding and national funding. The funding of rural projects all over the Europe became quite a massive operation, indicating a 'step of evolution' from welfare state Europe towards project state Europe.

In the context of European rural policy, a project is a short-term organisation, which aims to achieve long-term effects. This paradoxical setting may, in the coming years, give birth to new institutions, with an aim of complementing short-term projects with more stable backup structures and organisations. The dilemma is that such organisations may jeopardise the original idea of LEADER, that is, of moving the initiative in development matters down to the local level, in accordance with, for example, the subsidiarity principle. However, today, national state administrations, private consultant companies, and regional networks are competing with Local action groups for the last word in rural development.

Knowledge

One of the basic aims of LEADER is to sustain local know-how and implementations of tacit knowledge. In the aggregative setting the form of knowledge in an organisation is mostly formal and based upon expertise. When integrative elements are stronger, various types of knowledge co-exist. And as Munck af Rosenschöld and Löyhkö (Chapter 3) note, in such settings there are also attempts to integrate expert and lay knowledge. The practices of rural policy are constructed along the latter line. In this respect, rural policy is also learning from the experiences of action research, which supported combinations of research and practice, where researchers participate and take active roles on the one hand studying, and on the other teaching practical workers how to study their own work and environment (see more in Chapter 7). Such a line of action aims to change the state of affairs in social settings, in a similar way as LEADER is trying to change local circumstances in order to favour the type of development which local actors find favourable. According to this thinking, an increase of knowledge about the circumstances may lead to an increase of social capital as well as the creation of empowerment processes. If this succeeds, then not only would knowledge be produced, but also new abilities to create knowledge.

The Hungarian projects studied by Asztalos Morell (Chapter 11) reveal how many-sided a concept know-how is in practise. The preparation of a project application presupposes know-how about the application process, as well as know-how about the specific field of application. For example, the realisation of a bioenergy project presupposes technological know-how on renewable energy, know-how on entrepreneurial activities and know-how on the working of community development. Successful cooperation between different actors at the local and regional level is really needed in order to be able to solve such a set of questions and to realise a capacity-building process.

Is LEADER Enough?

The idealised picture of local development is often not so bright when the processes are studied on the ground. Pooling internal and external resources; constructing networks; starting a project, learning and interacting in order to increase social capital and to facilitate empowerment of local actors; – all this will not always be enough to induce successful and sustainable activities that push local development forward. However, the situation is far from discouraging overall. Several results give evidence of positive local development and far more results confirm positive reception of LEADER's basic ideas among local actors. In the following the authors present results from experiences in Denmark, Finland, Great Britain, Hungary, Italy, Russia, Spain and Tunisia. We conclude the volume with a summary of the results from three angles: the importance of the history of democracy and the democratic traditions in the countries where LEADER has been implemented, the likelihood that LEADER will improve democracy in former, and current, authoritative societies and third whether the benefits of the programme are reasonably equally distributed, especially in countries marked by inequality. Finally, we hope this volume will give the reader a more comprehensive picture of this many-sided policy approach.

References

Andersson, K. and Kovách, I., 2010. *Lagging Behind or Leader in Local Democracy?* Helsinki: University of Helsinki.

Asztalos Morell, I., 1999. *Emancipations Dead-End Roads? Studies in the Formation and Development of the Hungarian Model for Agriculture and Gender (1956–1989)*. Uppsala: University of Uppsala.

Bang, H.P., 2003. *Governance as Social and Political Communication*. Manchester and New York: Manchester University Press.

Bock, B.B. and Derkzen, P., 2008. Barriers to Women's Participation in Rural Policy Making. In: Asztalos Morell, I. and Bock, B.B., eds. *Gender Regimes, Citizen Participation and Rural Restructuring*, Oxford: Elsevier, pp. 265–84.

European Commission 2006. *The LEADER approach – A basic guide*, Luxembourg: Office for Official Publications of the European Communities [online], Available at: http://ec.europa.eu/agriculture/publi/fact/leader/2006_en.pdf [Accessed 7 December 2014].

Goodwin, M., 1998. The Governance of Rural Areas: Some Emerging Research Issues and Agendas. *Journal of Rural Studies*, 14(1), pp. 5–12.

Heffen, O. van, Kickert W.J.M. and Thomassen, J.J.A., 2000. *Governance in Modern Society: Effects, Change and Formation of Government Institutions*. Dordrecht: Kluwer Academic Publishers.

Kearney, B., Boyle, G.E. and Walsh, J., 1994. *EU LEADER I Initiative in Ireland: Evaluation and Recommendations*. Dublin: Department of Agriculture, Food and Forestry.

Kersbergen van, K. and Waarden, F. van, 2004. 'Governance' as a bridge between disciplines: Cross-disciplinary inspiration regarding shifts in governance and problems of governability, accountability and legitimacy. *European Journal for Political Research*, 43(2), pp. 143–71.

Klijn, E-.H., Koppenjan, J.F.M. and Termeer, K., 1995. Managing networks in the public sector, a theoretical study of management strategies in policy networks. *Public Administration*, 73 (3), pp. 437–54.

Kovách I. and Kucrová, E., 2006. The project class in Central Europe: The Czech and Hungarian cases. *Sociologia Ruralis*, 46(1), pp. 3–21.

———, 2009. The Social Context of Project Proliferation – The Rise of a Project Class. *Journal of Environmental Policy & Planning*, 11(3), pp. 203–21.

Lázár, I., 2010. *Vidékfejlesztési stratégiák* és *a fenntarthatóság* [Rural Development Strategies and Sustainability]. Unpublished study presented at Corvinus Egyetem, Budapest.

March, J. and Olsen, J., 1989. *Rediscovering Institutions: The Organizational Basis of Politics*. New York: Free Press.

Marsden, T., 2008. *Sustainable Communities: New spaces for planning, participation and engagement*. Oxford: Elsevier.

Murdoch, J., 2006. Networking rurality: Emergent complexity in the countryside. In: Cloke, P., Mardsen, T. and Mooney, P.H., eds. *Handbook of rural studies*, London, Thousand Oaks and New Delhi: Sage Publications, pp. 171–84.

Rhodes, R.A.W., 1996. The new governance: Governing without government. *Political Studies*, XLIV, pp. 652–67.

Shortall, S., 2004. Social or economic goals, civic inclusion or exclusion? An analysis of rural development theory and practice. *Sociologia Ruralis*, 44(1), pp. 109–23.

Sorensen, E. and Torfing, J., 2003. Network politics, political capital, and democracy, *International Journal of Public Administration*, 26 (6), pp. 609–34.

Polányi, K.,1976. *Az archaikus társadalom* és *gazdasági szemlélet* [The *archaic society* and economic philosophy]. Budapest: Gondolot.

Putnam, R.D., 2000. *Bowling Alone: The Collapse and Revival of the American Community*. New York: Simon Schuster.

Rantala, K. and Sulkunen, P., eds, 2006. *Projektiyhteiskunnan kääntöpuolia* [Flip sides of project society]. Helsinki: Gaudeamus.

Ray, C., 2000. The EU LEADER Programme: Rural development laboratory. *Sociologia Ruralis*, 40(2), pp. 163–71.

Sjöblom, S., Andersson, K., Eklund, E. and Godenhjelm, S., eds, 2006. *Project proliferation and governance – The case of Finland*. Helsinki: Helsinki University Press.

Söderbaum, P., 2011. *Bortom BNP, Nationalekonomi och företagsekonomi för hållbar utveckling* [Beyond GDP, Economics and Business Administration for Sustainable Development]. Lund: Studentlitteratur.

Thuesen, A.A., 2010. Is LEADER Elitist or Inclusive? Composition of Danish LAG Boards in the 2007–2013 Rural Development and Fisheries Programmes. *Sociologia Ruralis*, 50 (1), pp. 31–45.

Chapter 2
LEADER and Local Democracy: A Comparison between Finland and the United Kingdom

Johan Munck af Rosenschöld and Johanna Löyhkö

Introduction

European Union's LEADER is often regarded as one of the central tools for bottom-up rural development (Ray 2000). As the notions of locality and interlinking of actors from various sectors and areas are in the limelight of LEADER (c.f. Saraceno 1999, 443), participation and social inclusion become critical issues. In this empirical chapter we will evaluate the democratic characteristics of LEADER on the local level in Finland[1] and the United Kingdom. We will here treat LEADER as consisting of Local Action Groups (LAGs), individual LEADER projects and the regulatory bodies controlling LEADER. Thus, we use the concept of a 'local LEADER system' to reflect the relationship between these three groups of organisations. By using this concept, we emphasise that in order to gain a wider understanding of the democratic characteristics of LEADER we need to broaden the scope beyond Local Action Groups and also study how they operate together with LEADER projects and regulatory bodies on the local level.

The chapter is structured as follows. We first present and discuss the elements of LEADER as a tool for rural governance and local participation. Next, we briefly describe how the LEADER system as a whole functions both in Finland and the United Kingdom. Then we present the theoretical background, where we focus on describing the features of aggregative and integrative democracy through seven criteria, based on a modification of March and Olsen (1989). This is followed by a presentation of the case studies and the method used in this study. Next, we concentrate on the analysis. We first describe the case studies in both countries separately and then we compare the models of the local LEADER system of both countries. Finally, we discuss the results and its implications for the local LEADER system in general.

1 On Finland, see also Nousiainen in Chapter 6 of this volume.

Rural Governance, LEADER and Local Democracy

The notion of governance has been widely utilised in order to conceptualise the ways in which many problems are solved in contemporary society. It is therefore hardly surprising that governance has a wide range of meanings and connotations. In terms of rural development, the concept of rural governance is often used (see e.g. Goodwin 1998; Marsden 1999). This reorganisation of rural development can be seen in relation to the European Union Structural Funds, with their emphasis on temporary programmes and projects, which have entailed institutional changes on the local level in terms of how decisions are made and how the money is allocated (Ward and McNicholas 1998; Godenhjelm et al. 2012). European rural development has in other words seen changes similar to other sectors, such as environmental policy and sustainable development (e.g. Jordan 2008).

LEADER is generally depicted as a manifestation of this reorganisation of rural development. With its emphasis on inclusive and adaptive characteristics, LEADER is an example of a governance-type development mechanism of the late-modern age set out to deal with contemporary problems (cf. Andersson 2009). It is argued that LEADER has a two-fold role. On the one hand, it is a means of allocating funding for rural development, and on the other, it is expected to bring about not only changes in terms of development in an economic sense, but also concerning cultural, environmental and social issues (Andersson and Kovách 2010). For example, Osti (2000) argues that in Italy the LEADER programme signifies a shift from a more hierarchy-based rural setting to one resembling a market-based approach to development.

The social elements of LEADER have been studied to some extent. Although acknowledged for their socially empowering characteristics, Geddes (2000) argues that European local partnerships, such as LEADER, have in general only had limited success in involving socially excluded populations. Likewise, Kovách and Kučerova (2009) see LEADER as embodying a new form of rural class, 'the project class', which possesses the necessary knowledge and experience needed for initiating and administering LEADER projects. Thus, the local community should not be viewed as a homogenous entity, rather greater attention should be given to citizens without a strong public role (Hayward et al. 2004). Nevertheless, Hayward and colleagues argue that 'non-participation' does not necessarily entail social exclusion or lack of legitimacy, as personal commitments can prohibit certain individuals from participating, while they at the same time endorse a certain project. In a similar manner, Shortall (2008) argues that individuals, who do not participate in European Union rural development programmes may be actively involved in other arrangements outside of the European Union institutions.

In her study of LEADER in Denmark, Thuesen (2009, 39; Chapter 8 in this volume) highlights that the composition of Danish Local Action Groups' boards is 'characterised by a biased representation in relation to gender, age, education, main occupation and native country'. The study shows that board members are generally men in their 50s, highly educated, self-employed and Danish-born.

Furthermore, the study indicates that the Danish Local Action Groups are closed in terms of external participation – i.e. the Local Action Groups do not involve a wide range of local citizens to any great extent. On a similar note, Shucksmith (2000) argues that in the United Kingdom, social exclusion has been a challenge for LEADER, as it is dominated by individuals with high degree of social capital.

However, there is currently less evidence of how LEADER is linked to local participation concerning both Local Action Groups and LEADER projects and especially how these two entities work together. Additionally, the democratic implications of the relationship between the regulatory authorities and Local Action Groups as well as projects are currently not sufficiently covered in the literature. By focusing on the interlinkages between these three sets of organisations, we argue that we can find additional insights into how LEADER works on the local level in terms of participation.

LEADER in Finland and the United Kingdom

In this section we will briefly present an overview of how LEADER functions in Finland and the United Kingdom. The idea here is to give a brief introduction to the national contexts in which the programme is implemented.

The LEADER method was introduced in Finland in 1995 during LEADER II when it was included in the preparation of the second rural policy programme. The programme was approved in 1996 and the actual implementation of LEADER started in 1997 with 22 Local Action Groups. In the following programme period LEADER+ 2000–2006, 25 Local Action Groups were approved (Manner-Suomen maaseudun kehittämisohjelma 2007–2013 2011; Hyyryläinen 2007). During the programme period 2007–2013 there were 55 Local Action Groups in Mainland Finland and one in the Åland Islands. The LEADER programme was included in the Rural Development Programme for Mainland Finland 2007–2013, where LEADER represents one of the four axes that realise the rural strategy. The LEADER method can in other words also be used to realise the other three axes (Manner-Suomen maaseudun kehittämisohjelma 2007–2013 2011).

As LEADER started in Finland it encouraged a similar type of programme called POMO (Programme of Rural Development Based on Local Initiative) that funded Finnish Local Action Groups in the programme periods 1997–1999 and 2000–2006. POMO was based on the principles of LEADER, but it was financed entirely with national funding. POMO financed 26 Local Action Groups during the first programme period and seven in the last programme period (Manner-Suomen maaseudun kehittämisohjelma 2007–2013 2011). POMO was seen as more independent than the EU funded LEADER because of a lower level of bureaucracy and more flexible local development plans (Isosuo 1999, 119–20).

Activities of the Local Action Groups are currently governed by the Ministry of Employment and the Economy and the Centres for Economic Development, Transport and the Environment (ELY centres). The role of the Ministry is to approve the local development plans of the Local Action Groups and the group itself as an association (Finnish Council of State 2007). Even though Local Action Groups have the power to decide which projects will be financed, the final confirmation about the funding is made by the ELY centre of the particular LAG area (Finnish Parliament 2006; TE-keskus 2012).

In the United Kingdom, having been a separate programme for rural development, LEADER was in the 2007–2013 programme period realised as part of the national Rural Development Programmes in England, Wales, Scotland and Northern Island. In other words, the LEADER method has been integrated to the rest of the rural development policy by being a distinctive way of implementing the goals set out in the rest of the axes.

The LEADER axis is managed by different bodies in each of the countries. In England it is DEFRA (the Department for Environment, Food and Rural Affairs), in Wales the Welsh Assembly Government (WAG), in Scotland the Scottish Executive and in Northern Ireland the Northern Ireland Department of Agriculture and Rural Development. In all countries the managing body is also directly in contact with the Local Action Groups. In England this body used to be the Regional Development Agencies, which are now abolished.

Research suggests that LEADER in the United Kingdom has traditionally been concerned with economic objectives, such as enhancing job creation and maintaining economic viability in rural areas, instead of 'capacity-building' and inclusion of 'excluded individuals' (Shucksmith 2000, 214). Marsden and Sonnino (2008, 425), furthermore, argue that the LEADER programme has been a good tool for promoting local identification and the development of local products.

Theoretical Framework

It becomes obvious that concepts such as democracy or democratic integration need to be contextualised in order to be used in a valuable way. In this chapter, we operationalise the concept of democracy using a modification of the distinction between *aggregative* and *integrative democracy* by March and Olsen (1989). Furthermore, we divide the two categories into seven separate criteria (for an overview see Andersson and Kovách 2010; Godenhjelm et al. 2012). These criteria are depicted in Table 2.1.

Table 2.1 The Criteria for Assessing the Democratic Characteristics of LEADER

Criteria	**Aggregative democracy**	**Integrative democracy**
Actors *Which actors are represented?*	Political and administrative domination	Stakeholder domination
Institutional linkage *How are the projects related to the permanent administrative structure?*	Strong linkage through representation and working routines	Weak linkage
Forms of participation *What forms of participation are utilised?*	Little variation in forms of participation; project work characterised by administrative routine	Various forms of participation; (panels hearings, public meetings)
Conflict resolution *How are conflicts solved?*	Bargaining	Deliberation
Types of knowledge *Are different types of knowledge included?*	Domination of administrative and expert knowledge	Strong inclusion of lay knowledge
Outcome *Are there clear procedures for transferring new knowledge to the permanent structure?*	Clear administrative procedures	Unclear procedures
Accountability *Is there a clear political mandate?*	Clear political mandate, politicians can be held accountable	Unclear political mandate. Accountable actors not specified

Source: Based on Godenhjelm et al. 2012, 67.

Actors – refers to who participates in Local Action Groups and LEADER projects. In an aggregative setting the actors typically include civil servants and individuals with close connections to the permanent public administration. From an integrative point of view, the participating actors consist of a wider range of individuals and organisations, also including public bodies.

Institutional linkage – highlights the institutional position of Local Action Groups and projects. From an aggregative perspective organisations are closely aligned with the rest of the political system. Conversely, in an integrative context, these organisations are more freely organised with only weak ties to the surrounding system.

Forms of participation – relates to the characteristics or pattern of participation. From an aggregative point of view, participation is limited to formal procedures, where citizens 'are heard' to improve the work of the organisation. In terms of integrative democracy, participation is less confined to formalities, rather the ways in which citizens are integrated into the LEADER system is varied and more ad-hoc.

Conflict resolution – refers to how conflicts are managed and decisions are made. In an aggregative model the emphasis is laid upon formal structures, where individual preferences are 'aggregated' to form a basis for advancement and decisions are made generally by voting. According to the integrative model, conflicts are not resolved in a predetermined manner, but through deliberation and discussion. Here, the focus is on finding solutions through interactive learning.

Forms of knowledge – in an aggregative setting the form of knowledge in an organisation is mostly formal and based upon expertise. From an integrative point of view, various types of knowledge, both expert and lay, co-exist. In this setting there are also attempts to integrate these knowledges.

Outcome – relates to the output of an organisation and how this output is later on utilised. In the aggregative model the outcome of an organisation is clearly defined and means of controlling this process exists. In contrast, from an integrative point of view, the outcome is less clear and the control is likewise less strict.

Accountability – whereas outcome refers to the result of the organisation/ project and its control, accountability highlights how responsibility is divided within the organisation. In an aggregative model, the chain of accountability is clear with clear roles for the individuals. From an integrative point of view, these roles are ambiguous.

It is obvious that these criteria are empirically more complex and less clear-cut than as presented here. The same applies to the distinction between aggregative and integrative democracy. Nevertheless, these criteria work as a valuable analytic tool to explore how the LEADER system works on the local level taking into account a sufficiently wide range of factors.

Data and Method

The data for this chapter is based on the research project 'LAGging behind or LEADER in Local Democracy?', financed by the Finnish Ministry of Agriculture and Forestry (2010–2012). The aim of the project was to examine the democratic implications of LEADER programme in four European countries: Finland, Hungary, United Kingdom and Romania. In this chapter we focus on Finland and the United Kingdom (see Chapters 4 and 10 for Hungary and Romania).

Three case studies were made in Finland between November 2010 and January 2012. 37 interviews were conducted with Local Action Groups, including both members of the board and staff and project representatives. The selected cases represent different geographical parts of Finland: South-West Finland, Eastern Finland and Northern Finland. The region in South-West Finland is an archipelago area, the region in Eastern Finland is a lake area and the region in Northern Finland concentrates on the forestry and industrial sectors. Nine projects (three projects per Local Action Group) were selected in the study. Eight of these projects were development projects and one an investment project. Most of the projects were administrated by an association or in some cases by a local business and a few were owned by a municipality. The themes of the projects varied broadly from cultural education to sports activities.

The data from the United Kingdom consists of interviews with representatives from three Local Action Groups – two in Wales and one in England. All in all 12 representatives from Local Action Groups and 4 representatives from LEADER projects were interviewed in a three-week period in the summer of 2011. Of the studied LEADER projects two were investment projects for private companies and two development projects. The aims of the projects included developing environmentally-friendly tourist opportunities, cultural events and the development of local products. The Local Action Groups and projects were selected according to geographical location, so that our study would take into account the varying characteristics of the local areas. The Local Action Groups included in this study are presented in Table 2.2.

Table 2.2 Presentation of Local Action Groups Included in Our Study

Local Action Group	Area covered	Office location
1. I samma båt	Länsi Turunmaa, South-West Finland	Parainen
2. RajuPusu-LEADER	Southern Savonia, Eastern Finland	Juva
3. Elävä Kainuu Leader	Kainuu, Northern Finland	Kajaani
4. Menter Mon	Anglesey, Wales	Llangefni
5. Glasu	Powys, Wales	Builth Wells
6. Somerset Levels and Moors	Eastern Somerset, South West England	Langport and Taunton

The interviews were conducted as semi-structured interviews covering seven different themes representing the seven criteria presented in Table 2.1. The interviews for the first two cases were done with a tandem interview method by the first and second author, while the rest of the cases were done by the first author. The interviews were analysed using qualitative content analysis. In the analysis we identified indications and representations of aggregative and integrative democracy in the interview transcripts and coded them into themes in line with the criteria presented in Table 2.1 (cf. Silverman 2004).

Analysis

Actors

In the studied Finnish cases the composition of the Local Action Group boards strictly followed the tripartite system. The boards consisted of one-third public officials of the community, one-third associations and local companies and one-third independent local inhabitants. This tripartite system was also planned carefully beforehand. In one of the Local Action Groups it was even ensured that the municipalities in the Local Action Group area are represented according to the tripartite system in the board. In this same group also the position of chairman of the board circulated between the communities in the area in alphabetical order. Many of the interviewees mentioned, however, that it was hard to find local people to participate in the board. The administrative personnel of the Local Action Groups usually consisted of an executive manager and a varying number of project secretaries. The actors in the Local Action Groups thus represented more of an integrative logic.

The Finnish projects involved various actors, but they usually represented local people in local organisations and businesses. In some cases also public officials participated. Projects were usually owned by an organisation or a business or in some cases by the municipality. Some of the projects involved only a few participants, but others a variety of different actors. The project manager was usually an employee from the organisation or the business that owned the project. Two of the studied nine projects included one or more public officials. The actors in the projects thus mostly resemble a hybrid logic of democracy.

In the United Kingdom the actors included in our selected Local Action Groups and LEADER projects ranged from public officials to individual local citizens. The 'institutional position' of the Local Action Group influenced considerably which types of actors are included. For example, one Local Action Group was entirely detached from the local government, where the employees are 'third sector' workers, whereas in another Local Action Group, which was part of the local government, the employees were civil servants. Thus, the actors involved within the Local Action Groups are varied.

Interestingly, the concept of 'tripartite' division of responsibility of the Local Action Group boards seemed to be much more freely understood in the United Kingdom than in Finland. In the selected Local Action Groups the board representatives were predominantly representing public authorities, local companies and associations. The three Local Action Group boards did not include 'independent' individuals, who were not representing a specific organisation. The interviewees argued that the tripartite system would be ideal, but that they were struggling with finding new members.

However, in terms of the individual British projects the actors involved can be described as integrative. The project managers were employees from small local companies and local associations with no or weak ties to the local government.

Likewise, the participation of civil servants in the projects was limited to one project and with no direct influence on the project work. The reason for this may be that in the selected cases two out of four were managed by a private company. It can be argued that these small companies had no significant incentives to include members from neither the local government nor other local organisations.

Thus, in terms of actors the United Kingdom LEADER system can be seen as hybrid between aggregative and integrative characteristics.

Institutional Linkages

The Local Action Groups in Finland were all local associations that have been established to implement LEADER. One of the closest actors linked to the Local Action Groups were the regional ELY centres. The role of the ELY centres is to control the work of the Local Action Groups, give the final word about the selection of the projects and finance the activities of the Local Action Groups. The relationship to the ELY centres varied between the studied Local Action Groups, in one it was described as ‘warm’, whereas in the other two cases it was said to be more of a burden. The Local Action Groups were also linked to local organisations and other Local Action Groups. For example, one Local Action Group had cooperated with other Local Action Groups nearby and they had had joint projects and meetings. The Local Action Groups were also engaged in wider networks. Two of them had had international contacts in forms of international projects (both LEADER and other projects). Institutional linkages can be described both as aggregative and integrative due to formal and informal contacts with different organisations.

The LEADER projects in Finland had both a close and strong linkage to the Local Action Groups, but they were not directly connected to the ELY centres apart from via the Local Action Groups. According to the project interviewees, the relationship to the Local Action Groups was good and functional. The high level of bureaucracy was a regular theme in the interviews (with both project and Local Action Group interviewees). Some of the interviewees in one of the Local Action Group said that they function as a facilitator in terms of administrative requirements on behalf of the project workers. Projects were primarily working on their own and were not involved with other projects in the area. Due to the bureaucracy involved, but also the comparatively freestanding role of the projects in terms of connections to other projects and organisations, the institutional linkages of the Finnish local LEADER system can be described as hybrid.

The institutional linkages of the British Local Action Groups present an interesting case. The three Local Action Groups differed in terms of links to the local government in that one was an entirely independent company, one was part of the local government and the last was situated within the local government but operated ‘at an arm’s length’. Amongst the interviewees, there appeared to be mixed opinions regarding how close the Local Action Groups should be to the local government, as some saw the close relationship as potentially decreasing the amount of flexibility in their work. All the Local Action Groups had a close relationship to the regulatory authority responsible for administrating LEADER – in Wales the Welsh Assembly Government and in England DEFRA (previously the Regional Development Agencies, now discontinued). The selected Local Action Groups can therefore be seen to represent a hybrid form of democracy in terms of institutional linkages.

The LEADER projects in the United Kingdom all shared a strong linkage to the Local Action Group. Many of the project interviewees argued that the relationship to the Local Action Group had been very good. One project had some links to other non-LEADER projects, but the rest were almost entirely operating on their own. The linkage to the regulatory authorities was seen as strong, as many interviewees complained about the high level of bureaucracy involved in the management of the project. Thus, while not explicitly connected, the regulatory agencies were arguably strong players in the local LEADER systems.

The institutional linkages in the United Kingdom can thus be depicted as hybrid. The connection to local governments and regulatory bodies was close, yet the Local Action Group and the projects have a great deal of independence.

Forms of Participation

Participation in the Finnish Local Action Groups was possible mainly in two ways; by becoming a member of the board or by applying for projects. However, it was also possible to get involved by becoming a member of the Local Action Group organisation, being involved with drawing out the local development plan or by participating in the administration of the Local Action Group. The interviewees also stated that outsiders could comment on the work of the Local Action Group anytime, for example, in village nights and by e-mail, but that this was in fact not common. The participation structure in the Local Action Group is based on an aggregative logic due to its mostly formal paths of participation.

Participation in the Finnish projects was possible by either taking part in the realisation of the project or participating in the project 'as an object'. The project interviewees agreed that it had been possible to participate also after the project had been initiated. Information about the projects could be received through local newspapers and the homepage of the Local Action Groups. Some projects had been presented in a marketing letter that has been sent to local businesses and organisations. One of the interviewees stated that due to the size of the Local Action Group area it can be a challenge to reach the local people to inform about the work relating to LEADER. It was generally possible to comment on the projects and some projects use social media like Facebook. Participation in the projects thus follows both aggregative and integrative logic.

The level of participation in the local LEADER systems in the United Kingdom was low or non-existent. In terms of the Local Action Groups, the interviewees argued that they only provided limited opportunities for external local citizens to participate in their work. One interviewee stated that while they did not provide more structural means of participation, they were nevertheless 'out in the field' interacting with the local community. Another Local Action Group had used questionnaires to gather feedback, but the use had been sporadic. Also, another interviewee argued that participation 'happens' when compiling and writing the development plan for the Local Action Group and continued that this was the only route for participation in the Local Action Group. However, the interviewee stated that local citizens can participate through funded projects. For many Local Action Group interviewees, the question about external participation in the Local Action Group appeared irrelevant or even non-beneficial for their work. Instead, they saw more potential in dealing with individuals and organisations, or 'communities of interest', which are familiar with the questions at hand.

Likewise, all project interviewees in the United Kingdom argued that the issue of participation is often irrelevant. This is perhaps understandable, as many projects were managed by private companies, who may not see participation as beneficial for their everyday work. Nevertheless, some interviewees argued that participation took place within the activities that the projects organised. Thus, participation was limited to the outputs that the project produced.

All in all, participation in the British LEADER system was limited. Often the whole question was regarded as irrelevant by referring to either the low interest of local citizens to participate or the usefulness of doing so in the first place. These findings are interesting, as they highlight that the LEADER system is not promoting participation itself, rather participation is happening through projects (when it comes to the individual project managers) and the outputs of the projects (i.e. by participation in activities and events).

Conflict Resolution

In Finland the decision-making in the board of the Local Action Groups was based on deliberation and discussion. According to the interviewees, voting was rarely used and conflicts seldom occurred. If consensus about the financing of a particular project could not be reached, the project application was left on the table for further examination and discussion. Conflict resolution is highly integrative, as it is based on deliberation and reaching consensus.

In the Finnish projects decisions were usually made by the project manager, a smaller project group/working group or the steering team of the project. The operational decisions were usually made by the project manager and bigger issues were discussed in the steering team or project group. Conflict resolution in the Finnish cases can therefore be seen as following an integrative logic. (See also Nousiainen in Chapter 6 of this volume)

In terms of the British Local Action Groups, the predominant means of decision-making and conflict resolution was deliberation. Furthermore, the interviewees argued that few conflicts had occurred in their work. Interestingly, while all emphasised the importance of discussions, the decision-making process were different in the three Local Action Groups. One of them followed a more flexible process when making decisions, while another used formal scoring systems and appraisals made by external actors. On the whole, one can argue that the Local Action Groups followed a predominantly integrative way of conflict resolution and decision-making.

In terms of the British projects, the resolution of conflicts was done in an informal manner. In the interviews the question about conflict resolution appeared at times to be highly irrelevant, as many of the projects were managed by one or two persons. Nevertheless, the interviews reveal that decisions were made informally and after deliberation (if more than one person is involved in the project).

Thus, conflict resolution in the United Kingdom follows an integrative logic. Conflicts were solved and decisions were made in an informal way with an emphasis on discussions and no voting mechanism or apparent bargaining taking place. Here, conflict resolution is linked to forms of participation, as it can be argued that conflict resolution becomes more integrative when little external participation in the projects exists. The question remains; what are the conflicts and issues that are not deliberated in the LEADER system?

Forms of Knowledge

The Local Action Groups in Finland used both expert and lay knowledge in their work. In the interviews, both ELY centres and the knowledge of the board members were referred to as expert knowledge. In contrast, local people were mainly seen as representing lay knowledge. The Local Action Groups sometimes used outside consultants, but this was not very common. The forms of knowledge within the Local Action Groups can be seen as a hybrid of aggregative and integrative logics.

In terms of the Finnish projects, both lay and expert knowledge were used. Expert knowledge could be found either inside or outside of the project. One project manager, however, emphasised that he did not use outside expert knowledge at all. Local knowledge was often seen as important for the projects to succeed. However, one interviewee noted that it was not necessarily clear how to separate between lay and expert knowledge in practice. Often the nature and aim of the project guided which forms of knowledge are needed. Forms of knowledge used in the projects can thus be seen as a hybrid.

Concerning the forms of knowledge within the British LEADER system, expert and lay knowledge were both present. The everyday work of the Local Action Groups was characterised by a balance of formal regulations and informal flexibility. The regulatory influence stemmed from the directives and rules of the LEADER framework and the majority of the interviewees complained about the increasing bureaucracy of LEADER. Furthermore, two out of three Local Action Groups had used external consultants in their work (one group can be said to have used them systematically), often for evaluating projects and developing their own work. Therefore, expert knowledge played an important role in the work of the Local Action Groups. Simultaneously, the everyday work of the Local Action Groups was also strongly linked to lay and local knowledge. Many interviewees argued that the Local Action Group staff and board themselves have considerable knowledge and that the projects had utilised this knowledge in their work. Furthermore, one Local Action Group emphasised the importance of their networks and the knowledge these possess. Also, the Local Action Groups made use of external local knowledge, for example when evaluating project proposals.

In terms of the British projects, the dominant form of knowledge can be characterised as lay. The projects organised their work almost exclusively in a free manner and had very seldom used external consultants. However, one interviewee argued that they exchanged knowledge with other entrepreneurs in the region and that this had proven to be important for their work. The projects also utilised more formal knowledge when dealing with project management and reporting and evaluating their work.

Conclusively, hybrid forms of knowledge are present in the British local LEADER system.

Outcome

The Finnish Local Action Groups perceived their objectives broadly, emphasising that their aim was to keep the area alive and appealing by developing the social and financial viabilities in the area. The interviewees saw the role of the Local Action Group in the area in several ways. Many said that the work of the Local Action Group differs from the other forms of development activities in the area and that it has a close link to the grassroots level. Some of the interviewees also regarded the role of the Local Action Group to be more of a financer, administrator and a consultative organisation. Also, the objectives of the projects were primarily to increase the livelihood of a specific part of the Local Action Group area as well as to strengthen a sense of community. Some projects had more practical objectives, for example to offer exercise opportunities for adults.

The Local Action Groups did not systematically evaluate themselves, but they did write final reports about the project activities and annual reports. Some Local Action Groups discussed in an informal manner the results of their work and funded projects. All in all, no systematic evaluation of the outcome of the Local Action Groups was done apart from the annual reports and final reports that are part of the compulsory bureaucratic procedure.

As part of the mandatory reporting, the Finnish projects evaluated themselves and delivered the evaluation to the Local Action Groups. Most of the projects were assessed by the project group or steering team by filling in indicator forms, writing interim and final reports as well as by discussing the project informally within the project group. Many interviewees noted that reporting the quantitative figures was important. However, one interviewee stated that the results of the projects can only be seen after a few years' time. The outcome of the Local Action Groups and the projects can thus be depicted as hybrid.

The outcome of the local LEADER system in the United Kingdom can be characterised as hybrid. In terms of the Local Action Groups, they reported quarterly to the regulatory authority and in two cases as well to the local government. In addition, out of the three Local Action Groups two of them used some form of self-evaluation. One group used both quantitative and qualitative evaluation measures. On a general level, formal means of controlling the Local Action Groups and the implementation of LEADER thus existed. However, while referring to the development strategy of the Local Action Group, the interviewees understood the objective of their work in broad terms. What the Local Action Groups had in common is a strong emphasis on economic and business development in their respective region.

The British LEADER projects followed a similar pattern. On the one hand, formal mechanisms of controlling the projects existed. These include reports that were submitted to the Local Action Group on a quarterly basis and the final report once the project finished. On the other hand, the project managers viewed the objective of their activities in a much broader manner. While the application process was mainly seen as inflexible and bureaucratic, the process of reporting to the Local Action Group was on the contrary generally regarded as acceptable and not requiring significant efforts.

Thus, while formal means of controlling LEADER activity existed both in the British Local Action Groups and projects, they were simultaneously coupled with a broader scope of everyday work.

Accountability

The board was the responsible actor of the strategic work of the Finnish Local Action Groups in general, whereas the administrative staff and especially the executive manager were in charge of the practical work of the Local Action Groups. The chain of accountability was also clear in the Finnish projects, due to the small size of the projects and few project workers. According to most of the project interviewees, the main responsible actor was the project manager, the steering team or the chairman of the association. One of the interviewees said that the whole project team and everyone involved in the project were responsible for the project. Due to the clear formal chain of accountability the Finnish local LEADER systems can be seen as aggregative concerning accountability.

The British LEADER system appeared to have a clear accountability structure in place. In terms of the Local Action Groups, the actors responsible for the works were clearly defined; the board was ultimately accountable for the Local Action Group work on a general level, while the manager was responsible for the day-to-day tasks. The institutional position of the Local Action Group also partly influenced the chain of accountability – it can be argued that Local Action Groups that are situated close to the local government had consequently a stronger political mandate compared to independent Local Action Groups. Interestingly, one interviewee argued that they are, in addition to their funders, accountable to the local community and local people in general.

Also the British LEADER projects had a clear structure of accountability. The selected projects were small in terms of number of actors involved in the management of the project, which arguably made the question of accountability more straightforward. Accountability became taken for granted – it seemed normal that the manager of the project was in charge of the work.

By and large, the notion of accountability in the British LEADER system appears to be following an aggregative logic.

Discussion

In this section we discuss our empirical findings from Finland and the United Kingdom. The results from the individual countries are portrayed in Table 2.3 according to the seven criteria used in the analysis. We focus specifically on some selected issues that arose from the analysis and help to illuminate the differences and similarities between the two countries.

Table 2.3 The Democratic Characteristics of the Finnish and British Local LEADER Systems

	Finland	United Kingdom
Actors	Hybrid	Hybrid
Institutional Linkages	Hybrid	Hybrid
Forms of Participation	Hybrid	Aggregative
Conflict Resolution	Integrative	Integrative
Forms of Knowledge	Hybrid	Hybrid
Outcome	Hybrid	Hybrid
Accountability	Aggregative	Aggregative

As can be seen in Table 2.3, many of the criteria depicting the distinction between aggregative and integrative democracy resulted in being of hybrid character. This indeed suggests that the two-dimensional model of aggregative and integrative democracy is perhaps less capable of recognising the cultural differences in Finland and the United Kingdom than expected. Using the results of the dual model to directly evaluate the general democratic characteristics of LEADER may therefore be inappropriate. However, as became evident in the previous section, this 'hybridity' reflects different features in Finland and the United Kingdom and can therefore shed light on important differences between how LEADER works in the two countries.

One of the greater differences between Finland and the United Kingdom in terms of how LEADER is implemented on the local level is the 'institutional position', or relationship to local governments, of the Local Action Group. In Finland Local Action Groups were ad-hoc organisations with the only purpose to administer LEADER on the local level. These organisations were institutionally decoupled from local governments, but the local governments in the Local Action Group area have a 1/3 representation on the boards. In Finland the notion of a tripartite division of representation of the board was also understood more strictly – the board is composed of equal numbers of civil servants, local associations as well as companies and individuals with no formal links to either local governments or associations. In the United Kingdom, the institutional position of the Local Action Groups varied from being fully independent organisations to being part of local government. Moreover, the composition of the board of the Local Action Group was more varied. For example in one of the cases in the United Kingdom, the rules of the Local Action Group stated that no more than 50 per cent of the board should be composed of civil servants.

Interestingly, the institutional position of the Local Action Group did not seem to influence the forms of participation or conflict resolution. What the Local Action Groups in Finland and the United Kingdom had in common is that they did not allow external participation in their work to any larger extent. If they did, it was based on an aggregative logic. This was also largely the case in the LEADER projects in both countries, although some projects in Finland emphasised two-way communication with local citizens. However, our findings suggest that while the local LEADER systems were largely closed in terms of external participation, the internal participation in the LEADER systems appeared to follow an integrative logic. In other words, within the LEADER systems the emphasis was on discussion and deliberation and no greater barriers of cooperation between the actors existed.

Here, a couple of important questions arise; is the fact that the local LEADER systems are integrative in terms of conflict resolution a result of aggregative forms of participation? Can the LEADER systems maintain an open and deliberative way of dealing with conflicts and making decisions by not including external voices? The issue thus becomes who in fact participates in the programme. Many interviewees in our study stated that the best way to participate in LEADER is through project funding. The problem with 'project-based participation' is that it limits the range of actors who are realistically able to apply and administer projects in the first place. Previous research has found that people engaged with LEADER usually already have a certain position in their local community and knowledge of working with project-based rural development (e.g. Shucksmith 2000; Kovách and Kučerova 2009; Thuesen 2009). Because of this, the question of forms of participation becomes crucial when discussing social inclusion – if no effective means of integrating external individuals to the local LEADER systems exist, it will be highly challenging to assure a balanced social representation of the local area.

A third issue that arose from the analysis was that the process of transferring knowledge from the temporary project organisations to the permanent organisation was fragmented in both countries. While means of reporting and evaluating projects did exist, these arguably lacked the ability to sufficiently embrace the wide range of knowledge produced in the projects. The standardised forms that were used when evaluating LEADER projects may be useful for reporting quantitative data, such as spending, but they are less capable of accounting for more qualitative data relating to, for example, long-term capacity building and empowerment of local people. In addition, a clear majority of the interviewees in Finland and the United Kingdom complained about rigid administrative constraints relating to the application phase and reporting.

Conclusion

In this chapter, we studied the democratic characteristics of Local Action Groups (LAGs) and LEADER projects in Finland and the United Kingdom. Our study suggests that the Local Action Groups in both countries were predominantly closed for external participation. From a democratic point of view this is problematic, as incorporating actors with less experience of project-based activities then becomes less likely to succeed. However, we also found that the work within the Local Action Groups was characterised by openness and deliberation. The internal openness of the Local Action Groups highlights that once an actor has 'gained entrance' the work appears to follow a largely integrative model of democracy. An interesting question for a future study would be to assess whether the internal openness of the Local Action Groups is a result of the exclusion of external actors or if this is a result of the LEADER method itself.

This chapter also highlights that focusing specifically on the individual projects gives us a broader understanding of how participation takes place in the LEADER systems. Our analysis suggests that while the Finnish and British projects varied in terms of level of participation in the input stage of the projects, many of the projects could through their activities in fact be seen as promoting new routes of participation. This chapter argues that if we fail to properly consider the outputs of the projects in terms of participation, we end up with a limited view of the different ways in which LEADER is linked to local democracy. We suggest that more studies need to be conducted addressing the output of LEADER projects and their long-term democratic implications.

References

Andersson, K., 2009. Orchestrating Regional Development Through Projects: The 'Innovation Paradox' in Rural Finland. *Journal of Environmental Policy & Planning*, 11(3), pp. 187–201.

Andersson, K. and Kovách, I., 2010. *Lagging Behind or Leader in Local Democracy? An Assessment of LEADER-type Development Projects as a Tool for Democratic Integration in the Contested Countryside*. Helsinki: Swedish School Of Social Science, University Of Helsinki.

Finnish Council of State 2007. Valtioneuvoston asetus maaseudun kehittämiseen liittyvien ohjelmien hallinnoinnista 634/2007 [Act of the Finnish Council of State on Administration of Rural Development Programmes 634/2007], Helsinki: Finnish parliament [oneline], Available at: <https://www.finlex.fi/fi/laki/alkup/2007/20070634> [Accessed 7 December 2014].

Finnish Parliament 2006. Laki maaseudun kehittämiseen liittyvien ohjelmien hallinnoinnista 532/2006 [Act of Administration of Rural Development Programmes 532/2006], Helsinki: Finnish parliament [oneline], Available at: <www.finlex.fi/fi/laki/alkup/2006/20060532> [Accessed 7 December 2014].

Geddes, M., 2000. Tackling Social Exclusion in the European Union? The Limits to the New Orthodoxy of Local Partnership. *International Journal of Urban and Regional Research*, 24(4), pp. 782–800.

Godenhjelm, S., Munck af Rosenschöld, J., Kuokkanen, K., Andersson, K. and Sjöblom, S., 2012. The Democratic Implications of Project Organisations – A Case Study of LEADER-projects in Finland. In: Sjöblom, S., Andersson, K., Marsden, T. and Skerratt, S., eds. *Short-termism and Sustainability: Changing Time-frames in Spatial Policy Interventions*, Farnham: Ashgate, pp. 55–80.

Goodwin, M., 1998. The Governance of Rural Areas: Some Emerging Research Issues and Agendas. *Journal of Rural Studies*, 14(1), pp. 5–12.

Hayward, C., Simpson, L. and Wood, L., 2004. Still Left in the Cold: Problematising Participatory Research and Development. *Sociologia Ruralis*, 44(1), pp. 95–108.

Hyyryläinen, T., 2007. Toimintaryhmätyö paikallisen kehittämisen metodina [Work of the local actions groups as a method for local development]. *Maaseudun uusi aika*, 3/2007, pp. 20–36.

Isosuo, T., 1999. LEADER Leads to POMO. *Maaseudun uusi aika*, 2/1999, pp. 119–22.

Jordan, A., 2008. The Governance of Sustainable Development: Taking Stock and Looking Forwards. *Environment and Planning C: Government and Policy*, 26(1), pp. 17–33.

Kovách, I. and Kučerova, E., 2009. The Social Context of Project Proliferation – The Rise of a Project Class. *Journal of Environmental Policy & Planning*, 11(3), pp. 203–21.

Manner-Suomen maaseudun kehittämisohjelma 2007–2013 [Rural development programme for Mainland Finland 2007–2013], 2011 [Online]. Available at: <http://www.maaseutu.fi/attachments/newfolder_0/63DSiGFVB/Rural_Development_Programme_for_Mainland_Finland_041111.pdf> [Accessed 10 April 2012].

March, J. and Olsen, J., 1989. *Rediscovering Institutions: the Organizational Basis of Politics*. New York: Free Press.

Marsden, T., 1999. Rural Futures: The Consumption Countryside and its Regulation. *Sociologia Ruralis*, 39(4), pp. 501–20.

Marsden, T. and Sonnino, R., 2008. Rural Development and the Regional State: Denying Multifunctional Agriculture in the UK. *Journal of Rural Studies*, 4(4), pp. 422–31.

Osti, G., 2000. LEADER and Partnerships: The Case of Italy. *Sociologia Ruralis*, 40(2), pp. 172–80.

Ray, C., 2000. Editorial. The EU LEADER Programme: Rural Development Laboratory. *Sociologia Ruralis*, 40(2), pp. 163–71.

Saraceno, E., 1999.The Evaluation of Local Policy Making in Europe. Learning from the LEADER Community Initiative. *Evaluation*, 5(4), pp. 439–57.

Silverman, D., 2004. *Interpreting Qualitative Data: Methods for Analysing Talk, Text and Interaction*. London: Sage.

Shortall, S., 2008. Are Rural Development Programmes Socially Inclusive? Social Inclusion, Civic Engagement, Participation and Social Capital: Exploring the Differences. *Journal of Rural Studies*, 24(4), pp. 450–57.

Shucksmith, M., 2000. Endogenous Development, Social Capital and Social Inclusion: Perspectives from LEADER in the UK. *Sociologia Ruralis*, 40(2), pp. 208–18.

TE-keskus, 2012. Toimintaryhmät [Local action groups] [Online]. Available at: <http://www.te-keskus.fi/Public/?nodeid=10663&area=7543> [Accessed 31 May 2012].

Thuesen, A., 2009. Is LEADER Elitist or Inclusive? Composition of Danish LAG Boards in the 2007–2013 Rural Development and Fisheries Programmes. *Sociologia Ruralis*, 50(1), pp. 31–45.

Ward, N. and McNicholas, K., 1998. Reconfiguring Rural Development in the UK: Objective 5b and the New Rural Governance. *Journal of Rural Studies*, 14(1), pp. 27–39.

Chapter 3

A Perspective of LEADER Method in Spain Based on the Analysis of Local Action Groups

Javier Esparcia, Jaime Escribano and Almudena Buciega

Introduction

LEADER has become an instrument for the socio-economic development of rural areas in Spain. The ultimate goals of the European Commission's Initiative, launched in 1991, however, were to go further. Through innovative approaches such as Local Action Groups (LAGs), it was intended that LEADER would also be a tool for the empowerment of local society and social cohesion. The LAG is therefore conceived as an element in participatory democracy that in many senses was ahead of its time especially if we bear in mind other territorial (urban) contexts at that time.

In the following, we conduct a review of the LAGs in Spain from a qualitative perspective, from 1991 to the present. We provide an overview of the growing importance of the LEADER approach throughout this time period of over 20 years. The chapter also examines how from the early stages one of the main instruments of LEADER (LAGs) has had to maintain a difficult balance between two opposing forces. On one hand are the opportunities arising from the guidelines, set by the European Commission, of participatory democracy (new governance) and empowerment of local society. On the other hand are the trends arising from the conception of LEADER and LAGs as instruments of power controlled by local elites, and in some cases (more so at certain stages and in certain regions) in connivance with regional governments. At different scales, these local or regional elites have used LEADER and LAGs as clientelistic tools.

This chapter has two main issues. The first concerns the LAGs either as instruments of participatory democracy and public – private cooperation (governance) or as instruments of power in the hands of elites. And the second concerns the constraints (positive and negative) arising from the social and institutional environment in which LAGs develop their work.

The chapter concludes with some thoughts on a range of strategic issues, following two decades of experience, which could be considered in order to improve the effectiveness of LAGs so that they can continue to play a central role in the social and economic development of rural areas.

Theoretical Perspectives on LEADER between Governance and Power

It is known that LEADER marked two major innovations (Ray 2000; Shucksmith and Shortall 2001; Dargan and Shucksmith 2008; Shortall 2008). First, it is a territorial approach 'by and for' the local population, being a factor in the empowerment of local society and a means to design and implement strategies and actions in rural areas from a bottom-up perspective. Second, LEADER provides a tool for performing such tasks, the LAGs. LEADER performs a double function. First, it encourages (new) governance for rural areas (Goodwin 1998; Marsden and Murdoch 1998), providing a learning and capacity-building process for the local society and its most representative or dynamic actors. Second, it stimulates the democratisation of decision-making processes in local socio-economic development, which were previously controlled almost exclusively by public actors. In this context LEADER contributes to the legitimacy of collaborative stakeholder decisions (Connelly et al. 2006). However, this conception of LEADER has not always been predominant from the perspective of all rural actors. LEADER has also been the stage of tensions and power struggles between various elites, who have turned the programme in general, and LAGs in particular, into instruments of power.

In this section, we will use the literature to form a more in-depth analysis of these two main views of LEADER and LAGs (as the scenario for new rural governance and democratisation processes, and as an instrument of power and the stage of local elite power relations), in order to provide the context in which to analyse and raise the key issues for the Spanish case study.

LEADER as a Form of Rural Governance and Democratisation Processes

It is certain that LEADER has made significant advances in fostering governance networks, which may be an instrument for local democratisation (Hajer and Wagenaar 2003; Thuesen 2010). Governance, democratisation and emergence of networks are three key issues with which to better understand LEADER.

Most scholars concur that LEADER is an interesting attempt to implement a new form of governance in rural areas (Moyano 2001; Garrido and Moyano 2002). Some general key issues of new governance (Hajer and Wagenaar 2003; Rhodes 1996 and 1997; Stoker 1998; Sorensen 2006) are fully valid for the territorial approach to rural development. In the analysis of rural areas we may take into consideration Stoker's main propositions on governance (Stoker 1998). He points out first that governance involves a complex set of institutions and actors that go beyond the municipal government; second, that it assumes the presence of networks of actors (public, private and social) that enjoy autonomy in decision-making; third, that it focuses on the identification of economic and social problems, beyond the boundaries that exist between administrations and organisations operating at a local scale; fourth, that it allows the identification of dependencies and power relations between the institutions, organisations and actors involved in the various actions (taking the proper decisions); and finally, that governance recognises that the ability to make decisions and implement actions does not reside solely in the legal authority of the public administrations, but also in the authority that derives from the leadership of the institutions and actors involved in the development processes. In this sense LEADER rhetoric meets Stoker's propositions on governance.

But LEADER may be analysed also from its programmatic design (Böcher 2008). At least four main principles are present in LEADER that may be highlighted, which are central issues for efficient governance, and their implementation is a necessary – but not sufficient – condition for successful programmes.

These aspects are the contribution to self-governance (based mainly – but not exclusively – on the expected implementation of endogenous development processes through LEADER strategic plans); the relatively high decision-making capacity of the local actors; the inter-sectorial cooperation through networks and partnerships, and finally – complementary to the last aspect –integration, understood as the need to take into account all sectors of the rural economy as well as the involvement of all stakeholders, that is, the need for (effective) partnerships (Storey 1999).

In this context, LAGs may play the strategic role of Marsden's 'reflexive governance platforms' (Marsden 2013) but also may provide a forum for partnership, networking and consensus building (Lee et al. 2005). Their effectiveness could be crucial for the success of the programmes, since competent networks of stakeholders (such as LAGs) are more able to identify innovative solutions to the various problems and needs faced by rural areas (Thuesen 2010). Moreover, the importance of networks and networking processes for rural development has also been pointed out in the literature (Lowe et al. 1995; Murdoch 2000; Esparcia 2014), also as a key factor for the increase of social capital (Almudena and Esparcia 2013).

As elements for a democratic process LAGs are in theory open to citizens, allowing them to participate in giving opinions, contributing to the diagnosis of problems and requirements, and in the design of development strategies (Ray 2000). The more open LAGs are and the higher the degree of citizen involvement is, the more democratic they are. The legitimacy of this new rural governance is not automatic, since the representativeness of social and private stakeholders may be open to discussion everywhere. As has been pointed out, however, legitimacy is continuously constructed through discursive processes and a complex mix of competing rationales (Connelly et al. 2006).

LEADER as Scenario for Power Relations

In spite of the highly positive aspects of LEADER related to rural governance, democracy, partnership and networks (including social capital), some observations must be considered from the perspective of the practical implementation of LEADER. It has also an initial democratic deficit, because a number of the

LAG members are not elected through a democratic procedure. In the same way, networks of governance, such as those derived from LEADER, are sometimes seen as undemocratic due to the delegation of decision-making power to public, private and civic stakeholders (Thuesen 2010).

Probably the most fruitful interpretation of the 'negative externalities' of LEADER comes from the consideration of power as a matter of social production, in the context of new rural governance. In this sense LEADER could be interpreted as the scene in which actors and institutions attempt to gain capacity to act by blending their resources, skills and purposes into a viable and sustainable partnership (Stone 1989, cited by Goodwin 1998). Sometimes this intended viable and sustainable partnership responds to a paternalistic tradition, which may explain the distribution of stakeholders in LEADER and its decision-making bodies (Goodwin 1998). Frequently, however, new governance mechanisms have the purpose to ensure the continued hegemony of (some) local elites (Kovách 2000; Kovách and Kucerova 2006). This objective may imply a tendency to involve (especially in the decision-making bodies) only the key actors belonging to or coming from specific elite groups (public, economic or civic, or a combination of these).

With regard to power relations in LEADER, we found three main types of discourses, firstly related to the representation of different stakeholder groups (young people, women, politicians, etc.); secondly connected to the assumption of LAG responsibilities (and power) in the face of national or regional governments; and thirdly concerning to the territorial distribution of power within the LEADER areas.

In relation to the discourse of representation, it is certainly common that some groups (such as women, farmers, and young people) are less interested, or entirely uninterested in being involved in local structures for territorial governance (Shortall 2008), or are not well enough organised for this (Thuesen 2010), despite the fact that EU guidelines prescribe and support the broad participation of these groups (Böcher 2008). In fact, it has also been questioned whether LEADER always contributes to the capacity-building of excluded individuals or groups, redistributing power to the less powerful (Shortall and Shucksmith 1998; Shucksmith 2000). On the contrary, some authors argue that there is a tendency to favour those who are already more powerful and better articulated, and who may construct obstacles to the inclusion (or real involvement) of new actors in the decision-making structures of LAGs (Thuesen 2010; Esparcia 2011). In those situations, inclusion, empowerment and wider involvement of stakeholders and those groups with marginal positions is needed, as it contributes to the avoidance of elitism by the political class (Storey 1999; Scott 2004; Marsden 2013).

The second type of discourse concerns distribution of power from regional or national governments and LEADER. As Böcher has pointed out, 'the ideal of the autonomous self-government of rural areas through rural partnerships and networks, which is the central idea of regional governance ... rarely takes place in practice in its ideal form' (2008, 383). In practical terms, the main conflicts are caused by the reluctance of some officers in regional or national governments to let local actors to take on responsibilities in public fund management limiting autonomous steering at LAG level.

The third discourse concerns the territorial distribution of power, which explains much of the tension and conflict at the scale of LEADER. Within the LEADER regions it is common that a small number of economically more dynamic municipalities tend to concentrate more resources and power compared to other municipalities. Therefore, territorial tensions are not rare within the LEADER regions. Moreover, since the actors from economic or civic sectors are often less involved in LEADER, territorial tensions within the regions tend to be primarily of political nature (Esparcia 2011).

An Overview of LEADER in Spain

The Rise and 'Success' Of LEADER Approach

During the 1990s, LEADER already meant a real change of mentality in Spanish disadvantaged rural areas. Although at the beginning it could be seen by some sectors as a simple programme to channel aid to the poorest rural areas, it was gradually understood that it could be a genuine tool for development (Esparcia 2000; Esparcia et al. 2000a). Rural stakeholders were aware that the development of rural areas implies the productive diversification and the promotion of complementarity of income, and LEADER was a – partial – instrument for this purpose. LEADER was also an instrument for local management of these processes of development, based on cooperation between social, economic and institutional actors. Awareness of this issue, however, took more time to arise, and even today, in some cases, local actors do not realise the full potential of LAGs. But in general LEADER has been in Spain a novel means of approaching the problems of the rural world, not least because the local actors, for the first time, have been protagonists in important decisions affecting the development of their territories. Table 3.1 shows the main features of the various programmes and issues of LEADER and PRODER (note 3 in the table), the twin programmes implemented between 1996 and 2006. From these figures, highlighting an obvious growth in the area and population covered, and the public invested funding, many officials and politicians talk about the 'success' of LEADER in Spain.

Table 3.1 Basic Data and Some Observations of the LEADER Approach in Spain (1991–2013)

	LEADER I[1] (1991–94)	LEADER II[2] (1994–99)	PRODER 1[3] (1996–99)	LEADER Plus[4, 5] (2000–6)	PRODER 2 (2000–06)[4, 5, 6]	LEADER-Axis 4[7] (2007–13)
LAGs[8]	53	133	101	145	162	264
Km^2 (*1,000) (and percentage of country's total area)	82 (16.2%)	226 (45.0%)	120 (4.8%)	251 (49.8%)	234 (46.4%)	448 (88.8%)
Inhabitants (Million) (and percentage of country's total population)	1.85 (4.8%)	4.7 (11.2%)	4.4 (10.0%)	5.9 (13.4%)	7.6 (17.0%)	12.4 (26.8%)
Inhabitants / Km^2	22.6	20.8	36.7	23.5	32.4	27.7
Budget (Million €)	387	1,364	791	1,794	828[9]	1,474

Notes:

[1] Predominance of the private sector (10 per cent of LAGs): higher dynamism.

[2] General context: consolidation of LEADER as instrument for development. Nevertheless, some LAGs became instruments of power and political control at local level.

[3] PRODER (Operational Programme of Development and Economic Diversification of Rural Areas). Twin programme of LEADER restricted to Objective 1 regions but implemented in areas with fewer socio-economic limitations. Was successful enhancing productive initiatives in the secondary and tertiary sectors, as well as promotion of rural heritage (Esparcia 2001 and 2003b; Esparcia and Noguera 2004; Esparcia 2006).

[4] Despite the limitation of public sector in decision boards of LAGs, some regional governments and LAGs ignored the bottom-up methodology and continued using LEADER and PRODER as clientelistic and power instruments.

[5] Private investments were very important. The projects were less innovative in comparison with the previous stage.

[6] Some regions allow implementation of LEADER and PRODER in the same territories, by the same LAG, enhancing their complementarities (Esparcia 2009). Galicia funded an additional rural development programme (AGADER) focused on the most deprived areas (10 LAGs, 4,948 km^2, 0.23 million inhabitants).

[7] Integration of LEADER as specific axis of rural development, working together mainly with axes 1 and 2. Moreover it was expected to become an instrument to improve the governance in rural areas. Both aspects have failed in Spain.

[8] Number of LAGs and other collective actors.

[9] Initial public budget only; no data for private budget, but it exceeds public figures.

Source: Own elaboration from several years, Ministry of Agriculture (Spain).

The implication of these programmes, however, requires some additional remarks, mainly related to the last period, 2007–2013. It stands out because the territorial development programmes were integrated as a specific axis of rural development. The significance of the Axis 4 – LEADER – laid not so much in the actions but in the ability to manage measures and actions in other axes of this rural development policy, becoming, in theory, an instrument for the improvement of governance in rural areas. Nevertheless, the reality was very different. There was no confluence and understanding between agricultural and rural development lobbies, the former being afraid of potential interference and convinced of a drawdown of funds by the LAGs. The period 2007–2013 can be regarded as a failure in the attempt to extend and improve governance in rural areas through the LEADER method in Spain.

Other related problems were still present and overcoming them provides a significant challenge for the period of 2014–2020. These problems contribute to explain some of the results we will see in the next sections related to the role of LAGs. First, in many rural areas people primarily perceive the symptoms of failure and exhaustion of the model (and actors), accentuated by the increasingly strong bureaucratisation of the process (Sáenz 2011). Second, skirting the rules, some regions have virtually eliminated the actual functionality of the LAG (despite the preponderance of private actors, with an average of 57 per cent in the country), leaving the decision-making bodies as mere ‘stone guests’, with the real decisions taken by the regional political and administrative structures. Third, the dismantlement of experienced managerial teams has also contributed to make the programmes into a mere instrument of power in the hands of these regional governments and, in some cases, local elites (as in the regions of Valencia and La Rioja, in which in addition, non-public stakeholders in the period 2007–2013 accounted for slightly more than one-third of the stakeholders on average). Fourth, there are certain examples of LAGs that were clearly created from particular policy options (or clientelistic networks), and/or very close monitoring from regional governments (Sáenz 2011, 82). Although this is not a widespread situation, it necessitates a major renovation.

From Governance to Power Relations and the Role of Lags

As we have seen, LEADER has had a major impact in Spain, in terms of geographical coverage and because of the funds that have been mobilised (Moyano 2005). A comprehensive recent study lists these achievements (Ministerio de Medio Ambiente y Medio Rural y Marino 2011). Important progress has been made in the two major dimensions LEADER. The first dimension referred to is the diversification of economic activities (with the launch of a number of initiatives supporting the fragile rural economy). The second dimension is that of LEADER as an element for the improvement of governance, social capital and social networks, and the empowerment of local society. None of this has been achieved, however, without tensions and power conflicts.

Our analysis of LEADER in Spain shows that power conflicts and tensions are frequent, and they are often latent in local society (Esparcia 2000; Esparcia et al. 2000a and 2000b; Esparcia 2011; Esparcia and Escribano 2011 and 2012). Such conflicts arise with a high intensity in connection of LEADER, perhaps because of the control of resources. LEADER is conceived as an instrument of power and, as a consequence, is the subject of power struggles between the different elites. Power is present in a set of situations in which actors may play different roles, and power relations may go from a tacit consensus (to maintain compromises and equilibrium while available resources are shared according to some – not necessarily written – rules), to outright confrontation. The former situations are widely represented, but they usually mask some degree of domination-dependence relations rather than a fair equilibrium and consensus among local actors and society in relation to strategies, their practical implementation and, mainly, the distribution of resources.

At times there are different groups of elites. For example, local politicians fear loss of decision-making power (democratically legitimated by public institutions; that is, government structures) in favour of LAGs (not democratically legitimated, despite being included in government structures). In this sense it is not uncommon that public representatives constitute and act as power elites even against civic and economic stakeholders. On other occasions, territorial coalitions (from the same municipalities) between public, social and/or economic actors are the main driving force. Power conflicts could arise because some regional governments are sceptical on the capacity of local actors and LAGs to properly conduct the necessary processes and manage resources, and because from the perspective of regional government officials, LAGs are not sufficiently legitimated from a democratic point of view, including as they do unelected members, and managing public funds, which is not seen as their proper function. From such a discourse, which has been widely encountered, we may understand why some regional governments have not encouraged the development of rural governance and the empowerment of local stakeholders.

In the following sections we explore some of these key factors for a better understanding of the role of Spanish LAGs, which have been caught between the rhetoric of LEADER as an instrument for rural governance, social networking, capacity building, and local empowerment (in the context of democratisation processes); and the tensions and conflicts arising from its practical implementation and daily management, in which power relations could highly limit those achievements.

In the context of Spain, possibly the most important innovation of the LEADER approach to rural development has been the presence of LAGs. In the classical scheme of the LEADER method, LAGs form an element that not only channels funding and encourages cooperation or networking, but also constitutes the practical implementation of two of the three key factors of development: the creation of an institutional context for local development and the leadership of local stakeholders (the third being the territory) (Ministerio de Medio Ambiente y Medio Rural y Marino 2011).

Aims and Tools of Analysis of Spanish Lags

The aim of the following section is to conduct a basic assessment of local action groups of LEADER in Spain, according to three key elements. They are the LAG as an instrument for governance, the social institutional environment of the LAG, and the thematic working groups as one of the main innovations fostering participation and involvement of stakeholders. The methodology focuses on two aspects:

1. Consultation of available documents about the LAGs in Spain, although the information is highly fragmented and incomplete both in terms of territorial coverage and the temporal perspective. In particular, we focus on official evaluations and other 'grey' information published by the LAGs themselves as well as scientific papers on the issue (Esparcia and Noguera 2003 and 2004; ENRD 2010a, 2010b and 2011).
2. Analysis and qualitative assessment from two focus group sessions organised for this objective and conducted in late 2011 (Esparcia 2011). The two focus group sessions were conducted with team leaders (in the management teams) and presidents of some LAGs, respectively.

The material for the following sections thus mainly comes from the analysis of the documents about LAGs in Spain and from the ideas collected during the focus group sessions. Additionally, we make use of subsequent personal interviews with some of the participants discussing the results obtained.

Particular attention will be paid to two main issues. The first one is related to whether LAGs have been effective instruments for governance (as may be expected from the rhetoric of LEADER), or have they been more dominated by power relations with clientelistic patterns. The second issue is related to the social and institutional environments in which LAGs should work, and to what extent these environments could condition LAGs' achievements, facilitating or hindering effective governance by local actors in decision-making bodies.

Local Action Groups: Between the Rhetoric and Practical Implementation

Local Action Groups as an Instrument for Governance

In the late 1980s in Spain, there were no instruments of inter-municipal cooperation beyond the associations of municipalities (*mancomunidades*), which were restricted to the management or delivery of certain common services, for instance urban solid waste collection and social services. In this context, the implementation of LEADER was an innovation since LAG constituted in most rural areas the first inter-municipal body allowing territorial cooperation (Esparcia 2000). It was practically the first time that neighbouring municipalities had a forum for discussion and debate about common problems and aspirations. Under LEADER these forums could create development strategies; the impetus behind these was more or less shared, but mainly arose from the forums. Therefore, LEADER and LAGs became catalysts and expressions of a feeling of cooperation. In this sense it is true that while the Agricultural Mountain Act of 1982 began to build some relationships between rural municipalities, such relations were not comparable to even the first LAGs. In this regard, the ability of LAGs has been developed in recent years as a forum for discussion of problems and the needs of rural areas, often beyond what were the limitations of LEADER.

The first LAGs, however, are not comparable to the current situation. Indeed, in those early days the presence of public institutions predominated, due to the difficulties in mobilising the often scarce private actors, or simply via the control that some councils imposed or attempted to impose over these new bodies. Far from being understood as an instrument of development from the local perspective, in some cases the LAG were primarily conceived for public actors and, obviously more implicit than explicit, as instruments of power and consolidation or development of new clientelistic structures at the local scale (Esparcia 2000 and 2001). This situation, which was present (though not a widespread situation) in the early stages, was fortunately diluted by LEADER II and it can now be said that most local actors – both public and private – understand, conceive and work with the perspective of LAGs as instruments of territorial development.

Although it has occurred slowly and with some difficulties and exceptions, at present we can conclude that private actors have reached a more relevant position than they had in the early stages, having overcome the 'inferiority complex' with respect to public actors. This has been due, first, to the rules introduced by the European Commission in limiting the participation of public actors in the LAGs' decision-making structures, but also to the maturation of the private actors themselves with regard to their social role in the development of rural areas and within the LAGs. But despite these advances, and the fact that the EU rules nominally set the participation of public actors at no more than 50 per cent in the decision-making bodies, it is recognised that for practical purposes the influence of public actors in the final decisions is still above that of their nominal presence. In fact, it is not exceptional that public actors appeal to their contribution to the funding in order to maintain their dominant position in the crucial decision-making.

In spite of mentioned difficulties, LEADER has provided a clear relationship between public and private actors in rural areas, the latter having complemented the view held by public actors on the area and its needs. However it should be noted that a stronger presence of different groups of actors – both public and private – in the LAG does not necessarily imply that they have a shared territorial development project. In fact, the usual situation is to find cases in which they have cooperated only for specific purposes and to develop agreements on priorities which are frequently of the short- and medium-term. What is still lacking are medium- to long-term strategies, articulated and shared by all members of the LAG. In short, there are still no wide development strategies and no culture of commitment on joint and shared territorial development projects.

With some frequency, lobbies (territorial or sectorial) and even some powerful municipalities transfer their confrontations and interests to the LAG decision-making body, not always responding to the general interest. At times confrontations have emerged between public and non-public actors because of the different perspective, different priorities and ways to address the area's needs, or because of how interventions should be implemented. In this context, dominant elites tend to develop more strategic and well-practised methods to manage power and power relations based on clientelism.

In this sense it is often painted as a significant indicator of progress that LAGs have developed a culture of seeking agreements and consensus. Local actors boast that decisions are taken without voting, but otherwise they make no reference to the non-formal framework of negotiations, covenants and agreements established outside of the meeting decisions, based on the correlation of forces and the capacity to influence each of the different actors, as in a classical actors' game based on individual power relations.

Certainly this culture of covenant-consensus has made it possible to avoid earlier quite common situations, in which decision-making was a complex process, riddled with tensions and confrontations. Before the formal decision-making process, non-formal agreements were frequent, but usually they were controlled by and oriented to the interests of powerful actors, especially those of a public nature. This was criticised as 'politicisation'. To cope with and limit arbitrariness, private and also some public actors tried to establish clear technical criteria for decision-making (assuming that this meant a supposed objectivity and neutrality in the process). But criteria, even those of a more technical nature, are not neutral (Forester 1989).

Through an important process of maturation, LAGs increasingly became an instrument of participation and capacity building in rural areas (in the sense pointed out by Shucksmith 2000). Social acceptance of LAG members increased. Economic actors were more present and better represented. But a set of negative aspects still characterises LAGs in Spain (some of them not directly under their own responsibility). First, in parallel with social acceptance, expectations increased even about issues in which LAGs have no capacity to do anything. This inability to respond to social expectations led to frustration and dissatisfaction among certain sectors of the population, who do not have a clear idea on the limitations and real role of LAGs. Second, changes in the municipalities and public representatives within the LAGs, after political elections, are considered (mainly by managerial teams and private and social actors) as an element of instability and loss of efficiency in decision-making bodies as well as a slowdown in the operation of the programme. Third, some social groups such as young people and women are still usually underrepresented, at least in the decision-making bodies. Taking into account that the presence of these minorities affects articulation as a group and contributes to better networking in the local society (Bartol and Zhang 2007), it would be strategic to pay special attention to this issue. Fourth, loss of LAG autonomy with respect to regional government has increased recently, with negative effects on the implementation of the LEADER approach. The LAGs no longer have the initial autonomy and capacity for direct control of funds. Administrative controls became larger and more stringent despite the repeatedly announced simplification by the European Commission. During the period 2007–2013, some LAGs had an almost accessory or secondary role with respect to the regional government, being a perversion of the basic philosophy and LEADER approach. Fortunately these cases are an exception in Spain as a whole, so in no way do they tarnish the good work that has led to the development of regional governments and LAGs in most of the country.

The Social and Institutional Environment of the Lags

The LAGs are conceived as comprising an instrument with which to facilitate the participation of public and private actors, but this work takes place in an institutional environment that has a fundamental role to the extent that the actual operation of the LAGs can be affected. In general, the LAGs in Spain have been consolidated as an instrument of cooperation between different institutions of the LEADER region, mainly public, but also private. However, as the LAGs have been consolidated and strengthened, they have also become partners for other institutions, bringing increasingly more legitimacy in the institutional environment. There are three explanatory elements for this growing legitimisation (Esparcia 2011). First, the establishment of a forum for the participation of stakeholders in the territory, actors who have a clear legitimacy (city councils) and others with a growing recognition (economic and social actors, especially when they are articulated by associations or organisations). Second, the stringency of the work which the LAGs have progressed in the management of the programmes of territorial development, under which have been launched concrete and tangible actions that have benefited the socio-economic structure of the territory, and are thus appreciated mostly by local society. And third, this growing legitimacy derives from the fact that some LAGs have been able to assume and address functions and activities beyond the management of the LEADER programme.

This has meant recognition in the institutional environment which has contributed to the legitimization of the LAGs. Ultimately, the role of and the good relationships with public and private institutions have resulted in a greater flexibility in the management of actions and a greater proximity to citizens in general, and the beneficiaries in particular. The legitimisation of the LAGs in the rural territories is something already consolidated in most cases. They have been accumulating and now maintain an important credit of legitimacy in their immediate – local – institutional environment. This is different, however, to the situation in the regional institutional environment, in which the LAGs have reached much lower levels of legitimacy (with the occasional exception of the department of the regional government dealing with rural development).

In any case, there are exceptions to these general trends, and there are weaknesses that threaten the prestige and legitimacy that has been gained since the LAGs started to operate in the 1990s. Some of the elements that are at the basis of these difficulties in the legitimisation of the LAGs include changes in the composition of the groups; increase in size; localisms; ignorance, distrust and excessive intervention from regional bodies; loss of efficiency in the revitalisation work that has occurred in these years; and, among others, the inadequacy of instruments and means available to deal with functions that go beyond LEADER.

Indeed, territorial reconfiguration and increase in size has often forced the re-composition of the relationships between stakeholders (primarily public) and also to some extent the method of work, while developing intense efforts of pedagogy with the new members of the LAGs. All of this has slowed or even deteriorated the involvement and embeddedness of the LAGs in their social and institutional environment. It has also contributed to the maintenance and exacerbation of localisms, often linked to a lack of leadership, maturity and even appropriate knowledge on the part of public officials of what is meant by the LAGs and LEADER as instruments of cooperation and development for the territory. These difficulties have been reproduced with the political changes resulting from local elections, but also in the initial moments in each of the different periods. In this context some confrontations and struggles for power have emerged, thus contributing to the loss of prestige and social and institutional legitimacy of the LAGs.

The legitimacy of LAGs faces other weaknesses. On one hand, LAGs tend to have a high degree of uncertainty. This is because many of them are engaged with other additional initiatives, without a clear delimitation of such tasks (especially in the decision-making bodies) and, above all, without necessary instruments and work teams to address additional initiatives. The LAGs are engaged with and launch initiatives depending on the resources available at a given time, but they lack a clear strategy or medium- and long-term initiatives. In this sense, it is essential that the clarification of functions occurs within the management bodies as well as in the immediate institutional environment, i.e. clarification of the role of the LAGs in territorial development and the complementarities and cooperation mechanisms which can be mobilised with different institutions and public and private actors in the territory. Certainly there are significant even if still insufficient advances in this relation. In a large part, the social and institutional legitimacy of LAGs will be conditioned by the strength, transparency, effectiveness and commitment of the stakeholders that shape them.

On the other hand, aforementioned bureaucratisation and growing weight of daily management limit efforts in more strategic areas. This is more important when the legitimacy of the LAGs is seriously limited due to the political confrontations, the localisms and the possible lack of harmony between (primarily public) institutions. If the LAGs lose effectiveness in the work of revitalisation and as forums for discussion and strategic planning, they lose their legitimacy, not only in this institutional context, but also in relation to the local society.

An additional obstacle in the legitimisation of the LAGs is the excessive interventionism of some regional governments. Indeed, from the first moments in LEADER I and LEADER II, some regional governments have had a certain lack of trust in relation to the novel approach of decentralised management. This distrust resulted in the strict monitoring of the tasks developed by LAGs, ensuring that they worked within the limits of the functions related to the management and execution of a budget in a series of actions and specific projects (Esparcia and Noguera 2003). Obviously it was also a way to keep power in the hands of regional government officers and politicians in charge. But as the programme has progressed, it has in some cases slowed or stopped the process of legitimisation of LAGs (especially during the periods of LEADER Plus, 2000–2006, and even more during LEADER-Axis 4, 2007–2013). The reasons are diverse and range from ignorance of the implications of the LEADER approach to consideration of it as an instrument of clientelism, resulting both in the imposition of certain boundaries and in conditioning of the composition and operation of LAGs.

In addition to the already excessive bureaucratisation of LEADER, some Spanish regional governments imposed a basically administrative conception of the programme, with the justification (or excuse) of higher efficiency, which in any case is not proven. The result has been that many LAGs have suffered the loss of social and economic revitalisation functions to the detriment of the basically daily administrative follow-up conception. In this context, it is easy to understand the immense damage that these processes are doing to the LEADER approach. That will probably be highlighted in some of the upcoming evaluations of rural development programmes (2007–2013).

Conclusions and Strategic Issues for the Near Future

According to Spanish experiences, LAGs are among the most important political innovations, being a basic and central ingredient in the territorial approach to rural development. Local and decentralised management and cooperation between public and private actors are the essence of the LEADER method, which is one of the best examples of new governance in the context of European Union. However, as seen in the previous sections, the Spanish LAGs are still moving between their theoretical role as an instrument of development and being means of power relations including clientelistic practices. This dilemma exists both locally and on the level of regional governments, limiting greatly the progress towards mentioned theoretical objectives.

In the early stages of the implementation of the LEADER method and LAGs, many difficulties were faced, because of lack of experience, the slow pace of operations, consequent delays in effective implementation, and the high level of intervention in and control of the process by local public officials. There were no malicious intentions in these interventions from the side of regional governments, but they lacked previous experience, and private actors were poorly organised, even to the point of being barely visible. Gradually the procedures were improved with an increasing participation of private and social sectors and with clearer roles, functions and initiatives to be fostered.

A series of functions has been assigned to LAGs: first, the direct management of public funds through the promotion of specific actions (in certain sectors or specific areas); second, to be an instrument to facilitate the participation of stakeholders in development efforts and to revitalise local societies; and third, to design (with the support of managerial and professional teams) 'comprehensive development strategies', which means local strategic plans. These plans work as the reference in which actions are framed (Esparcia 2003a and 2006).

LAGs have certainly made a great effort in social animation and economic activities, and have reached many remarkable results. However, these results have been achieved with minor resources and often in difficult circumstances: without sufficient training or necessary knowledge and skills to design and to implement properly the LEADER strategic plans. Achievements have often been reached first and foremost up to the intensive personal involvement and the high commitment by the LAGs' members.

Even if a large part of the progress depends on commitments of local actors, it is also necessary to reach deep commitment of the regional government. According to past experiences, the lack of such a commitment used to lead to failures in adopting any serious and comprehensive policy for territorial development in Rural Spain. Unfortunately, it seems still today that the level of conviction is low among the regional governments, that LAGs could work as a potential instrument of territorial development beyond what LEADER as such offers (Escribano and Esparcia 2012).

Based on past experience, Spanish LAG approach should be developed around three main axes: rural territory as the scope of development (surpassing localism); strategic view as the steering principle, expressed in local development strategy; and seeing LAG as one key instrument among others.

More practically, LAGs should take into account a set of key issues, only to mention some of them here. For example, clearer stakeholder involvement and participation of private actors in the decision-making structures should be supported, paying particular attention to young people and women. Also, greater commitment of LAG members should be encouraged. Where appropriate, the completion of this work will enhance the networking within LAGs and between the different groups: young people, entrepreneurs, artisans, retailers, etc., and last but not least among women. One can list several other needs for reforming LAGs: training of managerial teams and of LAG members (especially on strategic issues); using self-assessment as a learning mechanism, especially for managers and LAG members; developing strategic thinking for the territory; and designing and implementing development strategies of medium- and long-term perspectives.

However, efficient work by the LAGs to reach such goals is difficult to be achieved without a genuine simplification of administrative procedures. LAGs should be allowed a greater role in facing the administrative management of the programmes, as well as in the implementation of the development strategies.

One of the basic achievements of LAGs is networking. In doing so, they contribute to the creation and development of bonding and linking social capital in rural areas. Networking is not just important within LAGs or between LAGs and other institutions in the rural area but between different territories or institutions outside the rural area. Here LAGs could take an important role in establishing and developing linkages between them and their respective actors; for example business centres, chambers of commerce, and various departments in regional government (Buciega and Esparcia 2013).

In relation to the external sphere of LAGs, the role of regional governments is highly important in Spain. Regional governments should provide more flexibility for LAG operations, avoiding interferences in decision-making, respecting the bottom-up methodology, and promoting positive elements in the culture of territorial governance. They should find solutions for lacking resources, needed to achieve the planned objectives as well as facilitate coordination between LEADER approach and other policies or programmes of special incidence in rural areas.

Our research results among the members in Spanish LAGs resume their request to return to the model used in LEADER I and LEADER II, in terms of design and operation of the programmes, and also their respect to the roles and functions of the LAGs in local society. In that time they were not, however, aware of the new context and the important changes that the reform of CAP and its second pillar will mean in the coming years for territorial and rural development. These changes constitute a framework that new – and probably very different – LAGs should deeply consider.

References

Bartol, K.M., Zhang, X., 2007. Networks and leadership development: Building linkages for capacity acquisition and capital accrual. *Human Resource Management Review*, 17(4), pp. 388–401.

Böcher, M., 2008. Regional governance and rural development in Germany: the implementation of LEADER+. *Sociologia Ruralis*, 48(4), pp. 373–88.

Buciega, A., Esparcia, J., 2013. Desarrollo, territorio y capital social. Un análisis a partir de las dinámicas relacionales en el desarrollo rural. *REDES. Revista hispana para el análisis de redes sociales*, 24(1), 81–112.

Connelly, S., Richardson, T., Miles, T., 2006. Situated legitimacy: Deliberative arenas and the new rural governance. *Journal of Rural Studies*, 22, pp. 267–77.

Dargan, L., Shucksmith, M., 2008. LEADER and Innovation. *Sociologia Ruralis*, 48(3), pp. 274–91.

ENRD, 2010a. *Summary of the Extended Report on the Implementation of LEADER Approach.* European Network for Rural Development, Brussels, p. 24.

———, 2010b. *The Implementation of the bottom-up approach. Extended Report.* European Network for Rural Development, Brussels, p. 44.

———, 2011. *Supplementary Report on specific aspects of LEADER approach implementation.* European Network for Rural Development, Brussels, p. 37.

Escribano, J., Esparcia, J., 2012. Social Context, social capital and rural development in the region of Valencia (Spain). *XIII World Congress of Rural Sociology*, 29 July – 4 August, Lisbon.

Esparcia, J., 2000. The LEADER programme and the rise of rural development in Spain. *Sociologia Ruralis*, 40(2), pp. 200–207.

———, 2001. PRODER 2: Características y principales innovaciones. *Actualidad LEADER. Revista de Desarrollo Rural*, 16, pp. 10–13.

———, 2003a. Valoración y balance de LEADER II. La consolidación de un modelo. *Actualidad LEADER. Revista de Desarrollo Rural*, 21, pp. 8–15.

———, 2003b. Valoración y balance de PRODER. Versión original. *Actualidad LEADER. Revista de Desarrollo Rural*, 22, 16–19.

———, 2006. LEADER II y PRODER en el Desarrollo rural en España. *Estrategias territoriales de Desarrollo Rural.* Institución Fernando El Católico, Diputación de Zaragoza, Zaragoza, pp. 65–89.

———, 2009. La nueva estructura de la Política Europea de Desarrollo Rural 2007–2009. Innovaciones y Retos. In: Gómez Espín, J.M., Martínez Medina, R., eds. *Desarrollo Rural en el Siglo XXI: Nuevas orientaciones y territorios.* XIX. Coloquio de Geografía Rural, Murcia 2009, Universidad de Murcia, Servicio de Publicaciones, pp. 167–202.

———, 2011. Los Grupos de Acción Local. In: Ministerio de Medio Ambiente y Medio Rural y Marino. *LEADER en España (1991–2011). Una contribución activa al desarrollo rural*, Madrid: Red Rural Nacional, pp. 97–124.

———, 2014. Innovation and networks in rural áreas. An analysis from European innovative projects. *Journal of Rural Studies* 34, pp. 1–14.

Esparcia, J., Escribano, J., 2011. Desarrollo territorial y reforma de la PAC. In: Moyano, E., ed., 2011. *El Futuro de la PAC es nuestro Futuro.* Anuario: Fundación de Estudios Rurales, pp. 142–9.

———, 2012. Social Networks and Social Capital in Rural Development in Spain. *XIII World Congress of Rural Sociology*, 29 July–4 August, Lisbon.

Esparcia, J., Moseley, M., Noguera, J., 2000b. *Exploring rural development partnerships in Europe. An analysis of 330 local partnerships across eight EU countries.* Valencia: University of Valencia.

Esparcia, J., Noguera, J., 2003. *Ex-post evaluation of the Community Initiative LEADER II. Spain.* European Comisión – DG AGRI, 12/2002 – 09/2003.

———, 2004. *Study for the methods for and success of mainstreaming LEADER innovations and approach into rural development programmes. Spain (PRODER Programme).* European Comisión – DG AGRI, 09/2003 – 03/2004.

Esparcia, J., Noguera, J., Pitarch, M.D., 2000a. LEADER en España: desarrollo rural, poder, legitimación, aprendizaje y nuevas estructuras. *Documents d'Anàlisi Geogràfica*, 37, pp. 95–113.

Forester, J., 1989. *Planning in the Face of Power*. Berkeley: University of California Press.

Garrido, F., Moyano, E., 2002. Capital social y desarrollo en zonas rurales. *Revista Internacional de Sociología*, 33, sep-dic, pp. 67–96.

Goodwin, M., 1998. The Governance of Rural Areas: Some Emerging Research Issues and Agendas. *Journal of Rural Studies*, 14(1), pp. 5–12.

Hajer, M., Wagenaar, H., eds, 2003. *Deliberative policy analysis: understanding governance in network societies*, Cambridge: Cambridge University Press.

Kovách, I., 2000. LEADER, a new social order, and the central –and East-European countries'. *Sociologia Ruralis*, 40(2), pp. 181–9.

Kovách, I., Kucerova, E., 2006. The Project class in central Europe: the Czech and Hungarian cases. *Sociologia Ruralis*, 46(1), pp. 3–21.

Lee, J., Árnason, A., Nightingale, A., Shucksmith, M., 2005. Networking: Social Capital and Identities in European Rural Development. *Sociologia Ruralis*, 45(4), pp. 269–83.

Lowe, P., Murdoch, J., Ward, N., 1995. Networks in Rural Development: Beyond Exogenous and Endogenous Models. In: van Der Ploeg, J.D., ed. *Beyond modernization: the impact of endogenous rural development*, Assen: Van Gorcum, pp. 87–105.

Marsden, T., 2013. From post-productionism to reflexive governance: Contested transitions in securing more sustainable food futures. *Journal of Rural Studies*, 29, pp. 123–34.

Marsden, T., Murdoch, J., 1998. The shifting nature of rural governance and community participation. Editorial. *Journal of Rural Studies*, Special Issue, 14(1), pp. 1–4.

Ministerio de Medio Ambiente y Medio Rural y Marino, 2011. *LEADER en España (1991–2011). Una contribución activa al desarrollo rural*, Madrid: Red Rural Nacional, p. 348.

Moyano, E., 2001. El concepto de capital social y su utilidad para el análisis de las dinámicas del desarrollo. *Revista de Fomento Social*, 56(221), 35–63.

———, 2005. Social and economic changes and players in rural areas of the south of Europe. First version in: B. Kasimis and T. Marsden, eds, 2003. *Agricultural policies and rural development in Europe*, Aldershot: Ashgate.

Murdoch, J., 2000. Networks – A new paradigm of rural development? *Journal of Rural Studies*, 16(4), pp. 407–19.

Ray, C., 2000. The EU LEADER Programme: Rural Development Laboratory. *Sociologia Ruralis*, 40(2), pp. 163–71.

Rhodes, R., 1996. The new governance: Governing without government. *Political Studies*, XLIV, 652–67.

———, 1997. *Understanding Governance: Policy Networks, Reflexivity and Accountability*. Milton Keynes: Open University Press.

Sáenz, C., 2011. Dinamización. In: Ministerio de Medio Ambiente y Medio Rural y Marino. *LEADER en España (1991–2011). Una contribución activa al desarrollo rural*, Madrid: Red Rural Nacional, pp. 69–92.

Scott, M., 2004. Building institutional capacity in rural Northern Ireland: the role of partnership governance in the LEADER II programme. *Journal of Rural Studies*, 20(1), pp. 49–59.

Shortall, S., 2008. Are rural development programs socially inclusive? Social inclusion, civic engagement, participation and social capital: exploring the differences. *Journal of Rural Studies*, 24(4), pp. 1–8.

Shortall, S., Shucksmith, M., 1998. Integrated Rural Development: Issues Arising from the Scottish Experience. *European Planning Studies*, 6(1), pp. 73–88.

Shucksmith, M., 2000. Endogenous development, social capital and social inclusion: perspectives from LEADER in the UK. *Sociologia Ruralis*, 40(2) pp. 208–18.

Shucksmith, M., Shortall, S., 2001. Rural Development in practice: Issues Arising in Scotland and Northern Ireland. *Community Development Journal*, 36(2), pp. 122–33.

Sorensen, E., 2006. Meta-governance: the changing role of politicians in processes of democratic governance. *The American Review of Public Administration*, 36, pp. 98–114.

Stoker, G., 1998. *Governance as theory: Five propositions. International Social Science Journal*, 50(155), pp. 17–28.

Stone, C., 1989. *Regime Politics*, Kansas: University Press.

Storey, D., 1999. Issues of Integration, Participation and Empowerment in Rural Development: The Case of LEADER in the Republic of Ireland. *Journal of Rural Studies*, 15(3), pp. 307–15.

Thuesen, A.A., 2010. Is LEADER Elitist or Inclusive? Composition of Danish LAG Boards in the 2007–2013 Rural Development and Fisheries Programmes. *Sociologia Ruralis*, 50(1), pp. 31–45.

Chapter 4
The LEADER Programme in Hungary – Bottom-up Development with Top-down Control?[1]

Bernadett Csurgó and Imre Kovách

Introduction: LEADER in the New European Union Member States

Local democracy in the Central and Eastern European rural context has been driven by administrative reforms and the introduction of a new development policy, primarily the EU LEADER programme, in which actors have democratic access to development sources and control over planning and decision-making processes. The analyses focus on the complex social/economic and political/power aspects of LEADER. For example, Bruszt and Vedres (2013) offer three dimensions (associating, mobilising and politicising) to aid understanding of local developmental agencies. The implementation of LEADER in new European Union member states is discussed in the literature as an act of importing political models (Maurel 2008), and the transfer of a governance model. According to Maurel's study, the impact of funding projects is not clear. Elected local officials may have initiated the formation of LAGs, but project managers were the ones who influenced the conceptualisation of development plans. Weak local participation, interest groups monopolising access to grants, traditional political practises such as paternalism, and interventionism may have sustained top-down power over bottom-up development practices. Furmankiewicz's paper on LEADER in Poland (2009) emphasises that the government distrusts voluntary, non-profit organisations; local government often considers NGOs to be rivals. In Romania, the administration has had difficulties in completing the SAPARD programme (Sharman 2003). In Harghita county, 72 per cent of LAGs originated from micro-regional associations (Harghita County Council 2005). According to a report about the Romanian LEADER programme (Chapter 10 in this volume), and the World Bank document (World Bank 2008), the entire system is still over-centralised and fewer tasks and responsibilities are delegated to regional, sub-state institutional

1 This chapter was supported in the framework of TÁMOP 4.2.4. A/2-11-1-2012-0001, 'National Excellence Programme – Elaborating and operating an inland student and researcher personal support system' key project. The project was subsidised by the European Union and co-financed by the European Social Fund.

and private actors. Analysts point out that the decentralisation of competencies, tasks and responsibilities started after 2007 (with Romania's European Union membership), but the necessary financial means are still under the control of the central state authority (Dragos and Neamtu 2007).

The rise of the project class (Kovách and Kucerova 2006 and 2009) and the active participation of non-profit, civil associations in LEADER programmes has changed the composition of actors in rural development, but the 2007–2013 LEADER programme still sees national authorities as having a decisive role in terms of management, control and institutional mediation (Halamska and Maurel 2010). By analysing the Czech LEADER Plus, Hudeckova and Lostak (2010) point out the problems involved with participation and involvement. In their view, the LEADER approach cannot be successful until important rural actors (farmers and others) and relevant rural activities have been marginalised. Only 35 to 45 per cent of Czech LAGs involve agricultural producers who have participated in LEADER (Hudeckova and Lostak 2008). The LAGs were not able to instigate the type of programmes which would have met the aspirations of many significant rural actors. This is why the LAGs in the Czech Republic, or in the other new member states, have been lagging behind LEADER principles. Focusing on LEADER implementation in the Czech Republic, Mathieu and Marty (2010) found that mayors had a dominant position, and that there was an unbalanced proportion of entrepreneurs and NGO representatives, and also that local/municipal administration played a key role in LAGs. Aunapuu-Lents (2012) argues that, as a consequence of hidden mechanisms of power originating from the multi-position of local leaders (sometimes belonging to political parties), representatives of local administration were able to control the decision-making of LAGs. However, on the boards of Estonian LAGs, the three main sectors (administration, entrepreneurs, NGOs) are represented equally. Aunapuu-Lents emphasises that: 'Without a vital civil society in the new European Union Member States, regional/rural development may become an instrument of project-class formation and domination over development funds'(Aunapuu-Lents 2012, 4).

Maurel (2009) presents, with the example of nature-protection projects, the way that local-interest groups limit the emergence of bottom-up principles. Provisions are made for local participation after the decision-making has taken place. Maurel also compared the implementation of LEADER in three Central European countries: Poland, Hungary and the Czech Republic (2008). LEADER mobilised local elite groups: 'A young generation of local managers and design office consultants, mainly graduates with an understanding of English, has surfaced as an active vector in transmitting the new LEADER thinking. They have realised the value that European programmes represent for their future careers' (Maurel 2008, 15). Another outcome of a LEADER initiative concerns the new relationships between citizens and their elected representatives, and the rise of new power-relation networks (Csurgó et al. 2010).

Local democracy has been attracting attention, given the core question of rural development. In this chapter, we study the power relations, the bottom-up and top-down character of LEADER in three Hungarian regions. In the first section, we present the implementation of LEADER in Hungary. In the second, we introduce three case studies by focusing on the motives of the actors involved, their networks, and the power relations inside and around Local Action Groups. In the final part, we compare three local models. In the summary, we put forward some of the criticisms of the political management of LEADER, along with potential new cooperation models for local democracy.

LEADER in Hungary

In Hungary, the reception of the LEADER programme has been less enthusiastic than in other European Union member states. It is true that LEADER has only 10 to 12 years of history in Hungary and its implementation is still in a transitory phase arising from many unresolved or unanswered questions. From 2001 to 2004, there was a tentative LEADER programme; during 2004–2006, LEADER plus was managed; while presently, LAGs are working on the LEADER 4 programme.

After 1990, during the preparatory period for European Union accession, the state offices, in partnership with international associations and agencies, launched LEADER-type, bottom-up, micro-regional development programmes. The first activity took place in 1992, with the intercommunal cooperation of the PHARE Regional Development Programme in 12 micro-regions, and special Hungarian–USA training programmes for the management and planning of development programmes (1994–1999) for micro-region activists. After the institutionalisation of rural development in the Ministry of Agriculture and Rural Development in 1998, the implementation of SAPARD (Special Accession Programme for Agriculture and Rural Development) opened a new phase in the history of bottom-up rural development. A sub-programme of SAPARD initiated the involvement of local actors in the planning of rural development. This was a learning process of planning and co-operating for many local leaders, managers, and civil activists, although SAPARD did not support any of the planned programmes. In total, 192 micro-regions joined the programme (about two-thirds of the current micro-regions). By 2001, 144 operational programmes were completed, including 4,300 fully elaborated project plans. The SAPARD started with significant leeway in autumn 2002 (Csite 2005; Nemes 2005).

The Ministry of Agriculture and Rural Development set up a pilot LEADER programme in 2001 which, due to the slowness of central administration, slippage of acquaintances and over-centralisation, had limited success (Fazekas and Nemes 2005). In a pilot LEADER programme, 25 million HUF (Hungarian Forint) were given to 12 associations, and 270 projects received support. After the re-assignment of 199 micro-regions in 2004, the SAPARD micro-regional development programmes were renewed. Learning how to plan was the most important outcome of the SAPARD pilot development programme, which highlighted the inefficiency of agricultural and rural administration.

One year after the Hungarian European Union accession in 2005, a two-phase programme commenced (AVOP LEADER Plus). In the first phase, local communities participated in a training programme and the most successful took part in the second step. Application deadlines were in September and October 2006. The beneficiaries were contracted in 2007, and 186 potential LAGs submitted proposals. The committees distributed 6.4 billion HUF to 70 selected LAGs (Figure 4.1). This affected 1.5 million people living in 920 settlements. The LAGs received 3,600 applications, and of the finally designated 2,700 projects, one third of funds were allocated to tourism and the preservation of cultural heritage, and the remainder financed co-operation, local economic development and farming. On average, the contracting took place 250 to 300 days after the application deadline.

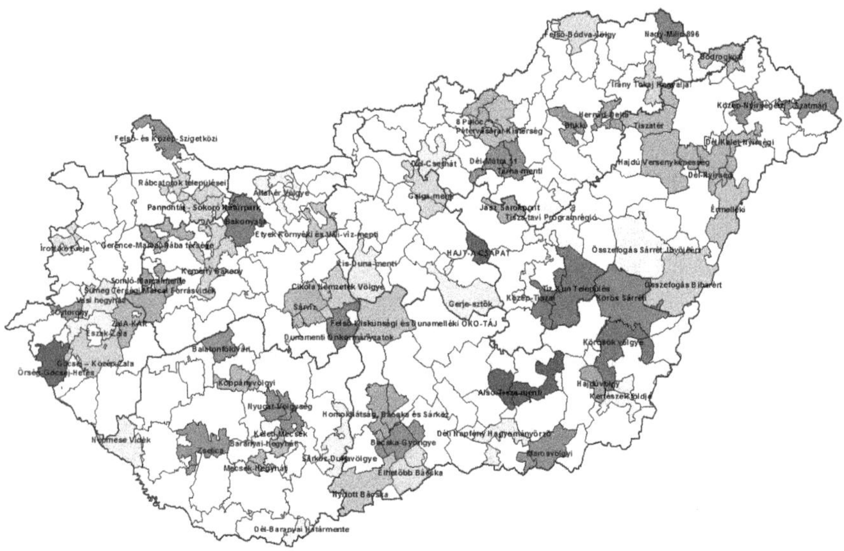

Figure 4.1 LEADER Local Action Groups in 2006

Source: Váti Kht.

The new phase of LEADER in Hungary started after the elaboration of the New Hungarian Rural Development Programme. In the spring and summer of 2007, local rural development groups began their activity. The political authorities initiated the enlargement of LEADER districts to 50,000 inhabitants. A total of 108 LAGs were founded by the end of 2007, although the New Hungarian Rural Development Programme calculated 150. The application procedure for the resources of the Third Pillar of Rural Development Plan started in 2008 October. The LAGs evaluated the proposals, but state offices oversaw the assessment procedures, and defined minimum scores which meant that LAGs could not carry local points.

The Agricultural and Rural Development Office has electronically registered and publicised the development objectives of 96 LAGs since 2009. The compulsory Delegation Contract between the Agricultural and Rural Development Office and LAGs permitted the receipt and evaluation of programme proposals. According to the evaluation of LAGs, the Delegation Contracts worked to dictate the conditions of cooperation, evaluation tasks, and the ranking of submitted project plans. An act declared that the Agricultural and Rural Development Office, on the basis of disposable sources and funds, would make the final decisions regarding financial support. The decision-making was long and complicated; applicants had to wait 10 to 11 months for information. LAG members, the local elite and experts, critically reflected on the long, state-controlled application procedure, and the over-bureaucratisation of the bottom-up LEADER programmes.

Compared to 2006, the permeation and territorial expansion of the LEADER Local Action Groups was considerable, as the map below suggests (Figure 4.2).

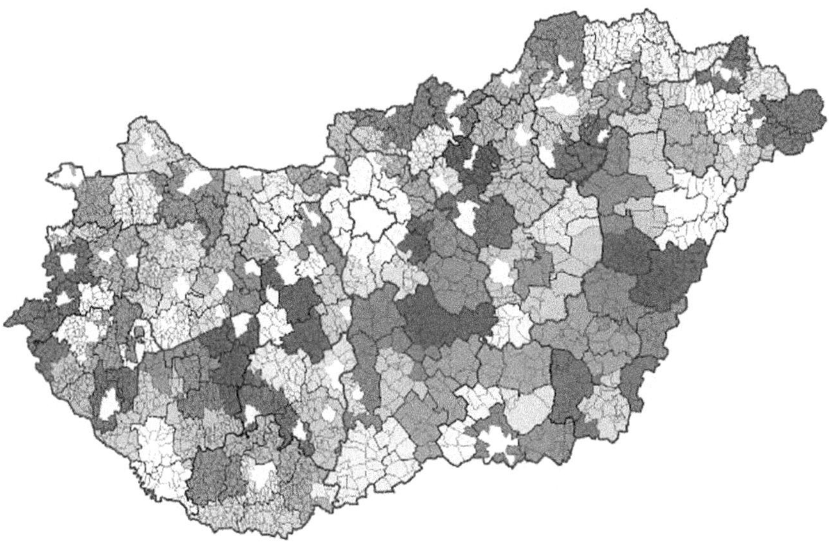

Figure 4.2 Municipalities Covered by the LEADER Local Action Groups (VKSZI)

Source: ÚMVP Report 2009.

It is true that the implementation of LEADER-type rural development programmes in Hungary is very inconsistent and conflicting. But, on other hand, we can agree with High and Nemes that: 'The interplay between institutions at different levels of governance can be seen in Hungary, for example, where accession to the European

Union has led to a significant growth in the influence of civil society by inducing the state to involve NGOs and civil organisations in decision-making processes' (High and Nemes 2007). Kiss and Szekeresné (2010) explain that capacity building can be the main benefit of the 2007–2013 LEADER Programme. According to Varga's study (2009) on the Southern Transdanubian Region LEADER Programmes, 37 per cent of the supported entities were municipalities or municipality-related institutions/companies, and 25 per cent were non-profit, civil associations. This chapter recognises the power-network nature of Local Action Groups (in relation to local democracy), and attempts to explain what kind of social and political motives work in LEADER projects.

Presentation of Cases

This chapter is based on three case-study analyses from different LEADER regions in Hungary. Between 15 and 20 interviews were conducted in each study area with members of Local Action Groups and with the relevant LEADER project actors. The case studies focus on the actors of the LAG, on the institutional linkages between the LAG and different institutions, on the forms of participation, on the decision-making process, on the forms of knowledge used in the LEADER system, and the regional specifics of LEADER outcomes. This case study analysis focuses on the type and amount of LAG members and the position of local government in relation to the LAG. As an indicator of the involvement of actors and their networks, we analyse the role of LAG personnel and the MVH (the Agricultural and Rural Development Agency) in all cases. The Agency consists of a central organisation operating on horizontal issues and directorates with official power, as well as county offices providing 19 representatives. MVH, as an institution, was founded to manage the applications for support, as well as for the allocation of support and the implementation of measures for the regulation of the market. It is financed by the European Union's agricultural and rural development resources and by the resources provided from the Hungarian national budget. The participation of locals in the project, and the characteristics of supported projects will be presented in order to demonstrate the decision-making processes, conflict resolution and knowledge use of the LEADER system in Hungary at the local level. Case studies will be analysed according to the analytical themes and related issues presented in Table 4.1. From the case-study areas, two LAGs are located in the eastern part of Hungary, both with different agricultural features, and the third in a tourist area in the region of Lake Balaton.

Table 4.1 Analytical Themes and Issues

Themes	Issues
Formation and composition of the LAG	Number of LAG members Position of local governments in relation to the LAG Main LAG actors
Involvement of actors and institutional relationships	Role of MVH Role of LAG personnel
Decision-making conflict resolution and knowledge use	Participation of locals in the project Characteristics of supported projects and knowledge use

The first case-study region is situated in the eastern part of Hungary, in Hajdú-Bihar County. The topography of Hajdú-Bihar County is characterised by the Great Plain. The LAG area includes seven settlements: four small towns and three villages. The area covers the main part of the historical Hajdúság region. The case of the Hajdú Competitiveness/Hajdú Versenyképesség (HAVER) Local Action Group highlights the multi-dimensionality of actors in rural development (Knickel and Renting 2000) and demonstrates several aspects of social-inclusion theory (Shucksmith 2000).[2] The case of the Nagykunságért Local Action Group region demonstrates how the LEADER, as the main form of local development, is generated by pre-existing local networks, and impacts on local community building. The case-study analysis refers to the theory of social inclusion (Shucksmith 2000), which presents the LEADER as a contributor not only to building local networks and community-related social capital, but also to promoting social inclusion.

The Nagykunságért LAG region is situated in the eastern-central part of Hungary, in Jász-Nagykun-Szolnok County. The LAG area includes eight settlements: five towns and three villages. The LAG region is located in the south-eastern part of Jász-Nagykun-Szolnok County. The most important sector of the economy is agriculture, which can be considered quite stable and productive. The population of the region is decreasing.[3]

The third case, the Dél-Balaton LAG, presents how the local project class acquires a central position in the local development system. As Kovách and Kucerova (2006) have proven, experts, designers, European and national

2 Interviews have been conducted with 19 different LAG members and representatives of LEADER projects for this case study. Three interviews were conducted with organisation staff of LAG; 13 were conducted with LAG members including local governments, civic organisations and entrepreneurs, and three were conducted with independent project leaders.

3 Interviews were conducted with 16 different LAG members and representatives of LEADER projects for this case study. Two interviews were conducted with organisation staff of LAG; 13 were conducted with LAG members including local governments, civic organisations and entrepreneurs, and one with an independent project leader.

administrative staff, holders of intellectual capital and representatives of the civil sector, along with other actors, occupy new social and project class positions. LAG staff (and other local institutions connected to the LAG) are occupied by local project-class members, and this has resulted in several conflicts and the re-structuring of the local power structure. The case-study region is in the western part of Hungary, in Somogy County. The LAG area includes 33 settlements, most of which are small villages; there are only five towns. The LAG area covers the larger part of the southern shore of Lake Balaton and a smaller region to the south-west of the lake. The LAG area is located in the region of Lake Balaton and its surroundings. Lake Balaton is already a proven tourist attraction: the most important economic sectors of this area are tourism and wine production. The LAG region is located in three micro-regions: some settlements from the Marcali micro region; the whole of the Lengyeltóti micro region; and the Fonyód micro region. Settlements within the Lengyeltóti and Marcali micro-regions are underdeveloped small villages characterised by a decreasing population and economic and social problems. The Fonyód micro-region is a popular tourism area.[4]

The Multi-Actor Development Network and Embryonic Democratism – the Case of the HAVER LAG

The formation and composition of a Local Action Group

The case of the HAVER Local Action Group proves the multi-dimensionality of actors in rural development (Knickel and Renting 2000) and aptly demonstrates several aspects of social-inclusion theory (Shucksmith 2000).

The HAVER LAG has continuity with the first LAG in the region, which worked within the framework of a local civic organisation named the Malomkő Association. In the second LEADER period (2007–2013), according to the national rules, all the supported LAGs had to create a formal organisation with the participation of all local LAG members. The HAVER LAG chose the form of a non-profit enterprise. It includes two different organisational parts: (1) personnel; and (2) LAG members separated into several subgroups and the Board. The LAG personnel include four types of employment status: one director, two project managers and one secretary. The LAGs can use 10 per cent of the support as a personnel cost.

Involvement in the LAG is based on the business share of the LAG enterprise. All members of the LAG have a different business share, which they bought when LAG was organised. There are 81 members in the HAVER LAG. According to the rule of the Hungarian LEADER system, 30 per cent of LAG members can be comprised of local government, and 30 per cent should be civic organisations, and 30 per cent should be economic enterprises. In the case of HAVER, there are 28

4 Interviews were conducted with 17 different LAG members and representatives of LEADER projects. Three interviews were done with organisation staff of LAG; 14 were with LAG members, including local governments, civic organisations and entrepreneurs.

civic organisations, 45 entrepreneurs and eight local governmental institutions, so the share of local governments is less than 10 per cent.

A total of 21 subgroups are active in the HAVER LAG area (for example, a subgroup of civic organisations for tourism in Ebes and another in the spa town Hajdúszoboszló, etc.). One subgroup has to have at least a 100,000 HUF business share. Sometimes, one institution – for example, a local government – can create a subgroup; but mostly, three or four organisations create a subgroup. All subgroups have an elected delegate on the board (six delegates from civic organisations, ten from entrepreneurs, and five from local governments). The share of local governments on the board is only 24 per cent. The board is the decision-making body. One of the subgroup delegates pointed out that he always has to stand for the interests and opinions of his subgroup; all the questions are discussed before board meetings: 'Sometimes I do not agree with my subgroup, but I have to vote according to their instructions'.

Involved actors in the HAVER LAG are mostly non-agricultural enterprises and local civic organisations. Finding members for the LAG was one of the main tasks for the personnel during the creation process. Most of the local actors thought that only the LAG members could apply for LEADER funds; it was the most important motivating factor for their participation. However, local actors can apply for LEADER funds without formal LAG membership. There are some local governments in the area supported by LEADER without LAG membership.

The involvement of actors and institutional linkages

MVH, in the view of local actors, is a really bureaucratic body which has a central position in the rural development project system. Many interviewees regarded MVH as slowing down the process, and most of them emphasised that the LAG seemed redundant and without a real function. The relationship between the LAG and MVH is bureaucratic and hierarchical. The LAG staff, as a local development body, intermediates between local actors and the MVH as a public body. LAG personnel do not comprise an independent development organisation.

The LAG has an important position in the local development system as an intermediate actor (Kovách and Kristóf 2009). The HAVER LAG is a good example of multi-actor rural development with a very strong bottom-up approach. Rural sociology offers classification and an actor-based analysis of rural activities (Tovey 1998; Kovách and Kucerova 2006). Kovách and Kucerova (2006 and 2009) argue that the theoretical categorisation of actors being involved in development projects is classified as a triangle of decision-makers/controllers, recipients of development funds, and goods and mediators between these two groups. LAG personnel engage with different actors in the region. MVH appears as the decision-maker/controller, while all the other local actors may be seen as recipients of development funds.

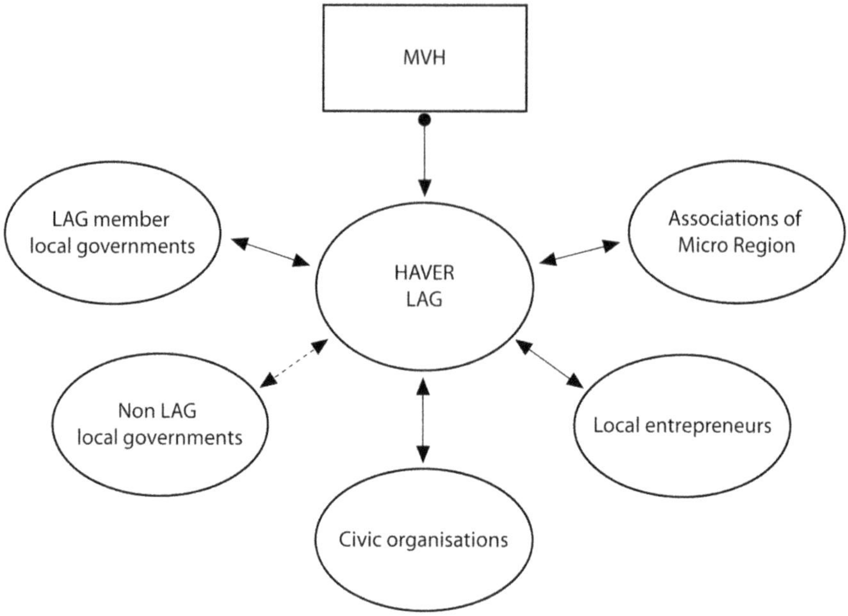

Figure 4.3 Institutional Links of the HAVER LAG

Local governments can participate in the LAGs as members if they buy business shares from the LAG organisation. Four local governments in the region bought business shares, but there are three which are not members of the LAG. The LAG-affiliated local governments have stronger connections with LAG personnel and also with other LAG members, participating in decision-making processes, and they apply for LEADER projects. Non-LAG-affiliated local governments have links with LAGs as applicants; the connection between that kind of local government and the LAG is very weak. These local governments feel that LEADER is less important in the development of their settlements, as well as in the region.

The HAVER LAG is linked to several local civic organisations. LAG personnel encourage them to participate and to apply for project funds. The most important civic organisations are the board-member organisations participating in decision-making. LAG personnel try to create connections with almost all the civic organisations of the region, and facilitate participation and application. The LAG also cooperates with local entrepreneurs. The most active entrepreneurs are members of the board, while others only appear as applicants.

Associations of the Micro Region are important development bodies in the region. The HAVER LAG covers two micro-region territories. These associations manage the common charges of local governments and appear as applicants for development projects. Personnel with these associations have the intellectual capital and experience needed for project management. Most staff members of the LAG have worked in this kind of association. There is a strong informal link between the LAG and associations, based on personal connections between the staff.

The institutional links of the LAG have two levels (Figure 4.3). Links with MVH are bureaucratic and hierarchical; LAG is under the control of MVH. At the local level, the LAG has informal links with different institutions, and the position of the LAG in the local-development system is horizontal. Relationships between the LAG and different local actors depend on the informal connections between them, and the level of involvement of LAG actors.

Decision-making, conflict resolution and knowledge use

The HAVER LAG successfully activated local society, resulting in the LAG being able to support 112 projects from 2007 until the end of 2010. The LAG used several forms of involvement. They organised information forums for locals to introduce the work of the LAG along with application possibilities, and they collected local plans and demands, allowing local citizens and institutions to give feedback on the work of the LAG. LAG personnel organised forums several times a year. Other important form of participation for locals has been a survey on project ideas and interests run by LAG personnel.

There are 81 members in the HAVER LAG. Members of LAG were divided into several subgroups and a Board with the main role of decision-making. The Board is comprised of subgroup delegates. Delegates stand in for subgroup members and their interests and demands. Delegates are elected by their subgroup members. There are six delegates from the civic organisations' subgroup, ten comprised of entrepreneurs and five from local governments. The main role of 21 subgroups is to handle negotiations; before any decision-making, they can send their opinions to the board on strategies, on projects, and on all the relevant LAG tasks. The Board decides on strategy and also on project support under the control of MVH. Before the decision-making process, the LAG personnel prepare all the documents according to MVH regulations. The project-application process, and also the administration process of the project, needs intensive cooperation with LAG personnel staff. Most of the project leaders emphasised the important role of LAG personnel, from project planning to the administrative closure of the projects. However, some of them stated that, because of the power of MVH, LEADER is too bureaucratic and complicated, and many local citizens are not able to meet the requirements. They highlighted that, because of MVH's position, the role of the LAG is only formal and without real power.

In order to prevent conflicts, LAG staff have organised meetings to negotiate planning. What is especially important is that they organised presentations of project ideas and plans. This resulted in most of the applicants' projects fitting in with the Local Development Programme, as well as the guidelines of LEADER in general.

Another potential time for conflicts is during the project application phase. The LEADER board decides on which projects can receive LEADER funds. Before this decision, LAG personnel discusses potential projects, assisting applicants to tailor their project to the formal regulations of LEADER. LAG personnel create a list of potential projects. After the local decision-making of the board, the LAG personnel sends the list of potential projects to the MVH, which controls all the project plans and makes the final decision on funded projects.

The main action of the HAVER LAG for preventing conflicts is the practice of 'supporting as many projects as possible and all settlements'. This means that they allocate a lower budget to a greater amount of projects; for example, in the case of rural tourism, the total support would be a maximum of 5 million HUF (€17,000). During the decision-making process, attention is paid to giving support to all the settlements of the LAG region. The most important actors of conflict prevention and resolution are those LAG personnel who are in contact with all the other actors. Their social networks determine the activities of local society. The LAG personnel also have direct contacts with MVH. Knowledge and central network positions of LAG personnel provide the basis for their central role among other actors.

We argue that paying attention to the networks of social relationships can help to understand the dynamics of social involvement and relationships between actors and capacities, as well as their place in the development process (Lee et al. 2005). According to the perceptions of informants, the question of accountability is not central to the thinking of the LAG. The LAG is responsible for its work and its activities. The LAG is responsible for the legal and administrative workings of LEADER in the region. On the other hand, as the informants stated, each local actor is responsible for its activities. Local governments are responsible for the development of their settlement, entrepreneurs for their own enterprises, and civic organisations for their activities. Local actors do not perceive that the LAG has overall responsibility for local development; it is only responsible for the short-term outcomes of LEADER through supported projects.

Most of the informants perceive that MVH has overall responsibility. It is responsible for the framework and rules of LEADER, and therefore, it is responsible for the long-term outcomes of LEADER-type development in Hungary. MVH supervises LEADER, and local informants believe that it has a moral and legal responsibility. Most of the informants have criticised LEADER, regarding it as too generalised and bureaucratic. As we described earlier, LAG's work and position in the LEADER system is thought to be only formal, without real power or responsibility.

According to Ray (2001), the LEADER system can be seen as a new form of governance which gives more power to local institutions. The replacement of local institutions involves a change in the way that knowledge is used for management. Local institutions tend to use their local, mostly tacit knowledge (Bruckmeier 2000). The shift of the knowledge system is one of the major impacts of local institutions on government methods because it is often accompanied by a change in control over resources. (Csurgó et al. 2008) The main form of knowledge used by the LAG personnel is managerial. The members of the LAG mostly use their lay knowledge. Because of the lack of managerial knowledge, most LAG members and potential applicants need the help and assistance of a professional project manager. There are some project managers locally who are aware of the LEADER system, and they mostly write and manage the local LEADER projects. Only some of the local project applicants use the assistance of external project managers without any local embeddedness.

Most supported projects are connected to rural tourism and rural heritage. Civic organisations, entrepreneurs, and also local governments have been active in rural-tourism projects. However, there were not enough applicants to develop the competitiveness of the Hajdú region; there was a form of development programme for local SMEs, but they were not active enough in the application process. So, the LAG aims to transfer resources from the budget towards tourism. This kind of decision needs the permission of the MVH.

The case of the HAVER LAG has demonstrated the multi-dimensionality of actors and their activities in rural development, and the success of social involvement through LEADER-type development. This demonstrates the relatively wide participation and multi-dimensionality of actors. HAVER leader groups have been innovative in this respect.

LEADER and the Local Oligarchy – the Case of the Nagykunságért LAG

Formation and composition of the LAG

The story of the Nagykunságért Local Action Group demonstrates how LEADER, as the main form of local development, can be generated by local networks and can have an impact on local community building. The case-study analysis refers to the theory of social inclusion. Shucksmith (2000) presents LEADER as a contributor not only to building local networks and community-related social capital, but also to promoting social inclusion. The current LAG organisation is strongly connected to the first LEADER organisation in the region; and in the new LEADER period (2007–2013), most of the previous members had joined the LAG. The current action group consists of eight settlements from the original ten. The legal form of the LEADER in the region is a flexible non-profit enterprise; if local actors want to join the LAG, they only have to buy shares in the business. The amount of support has become much higher; the LAG can obtain more subsidies for supporting local projects. In the first period of LEADER action, a local rural-development expert group and the LAG personnel made decisions without any formal control; the LAG was a pre-decision-making body, and the final decisions were made by MVH as the control organisation. The independence of the LAG organisation of the local government resulted in many problems. LEADER projects are post-financed, so the LEADER organisation needs credit to finance the cost of personnel. Nevertheless, the support could not cover the extra cost of credit. During the first period, the local government needed to manage the problems of post-financing.

The case of the Nagykunságért LAG illustrates how local notables dominate the process of rural development, referring to the social-exclusion characteristics of LEADER-type development (Shucksmith 2000; Shortall 2004). Exclusion is a much broader concept than poverty (Shucksmith and Chapman 1998), referring to the power of actors (Csurgó et al. 2008) and participation. Shucksmith argues that: 'the more articulate and powerful individuals and groups were better able to engage with programmes and to apply for grants and submit proposals, while others lacking the former's capacity to act were unable to benefit' (Shucksmith 2000, 210). Only those actors who have knowledge capital and networks are able to participate in the rural-development process (Csurgó et al. 2008).

The principal LAG actors are the LAG personnel, including the president of the LAG, two project managers and two project assistants. The president of the LAG is a local expert who was the leader of the LAG personnel in the first period when serving the post of local mayor. The importance of the LEADER is very high in local policy, as most of the local interviewees emphasised.

Involvement of actors and institutional networks

There are 39 members in the LAG with their own business shares. The share capital is 1,950,000 HUF (€7,000); six local governments are members of the LAG. In the case of the Nagykunságért LAG, only 15 per cent of LAG members are from local governments. The LAG board has 13 members, representing 13 subgroups of the LAG; each subgroup consists of 3 LAG members. The most active actors of the local LEADER are from Mezőtúr, which is one of the biggest towns in the LEADER region. The local government of Mezőtúr is a member of the LAG. The local government of Karcag (another large town) does not participate in the LAG. Local informants emphasised that Karcag wanted to create its own LAG, and when it was frustrated, the settlement joined the Nagykunságért LAG; however, local actors from Karcag are mostly unmotivated.

The villages in the region have a special position in the LEADER. In some cases, they receive positive discrimination as applicants; this means that there are special programmes and funds inside the LEADER and only village actors can apply. In addition, only the local governments are able to send applications from these villages; there are no local actors able to do so.

Agricultural enterprises (Figure 4.4.) were the most active actors in the region during 2004–2006; 55.8 per cent of projects financed by the European Union were connected to agriculture. Agriculture had a central position in future-development strategies and plans. Local agrarian actors had the knowledge and capacity to participate in development projects. In addition, they only ran individual projects, and did not take part in cooperation. Agriculture was one of the main topics of the first LEADER plan. According to the new regulations of LEADER 2007–2013, agriculture is excluded from LEADER funds. The rule is that firms making more than 51 per cent of their profits from agriculture may not apply for LEADER funds. Local actors re-wrote the LEADER development plan, focusing on local community building and the development of non-agricultural enterprises. Most active actors became non-agricultural enterprises.

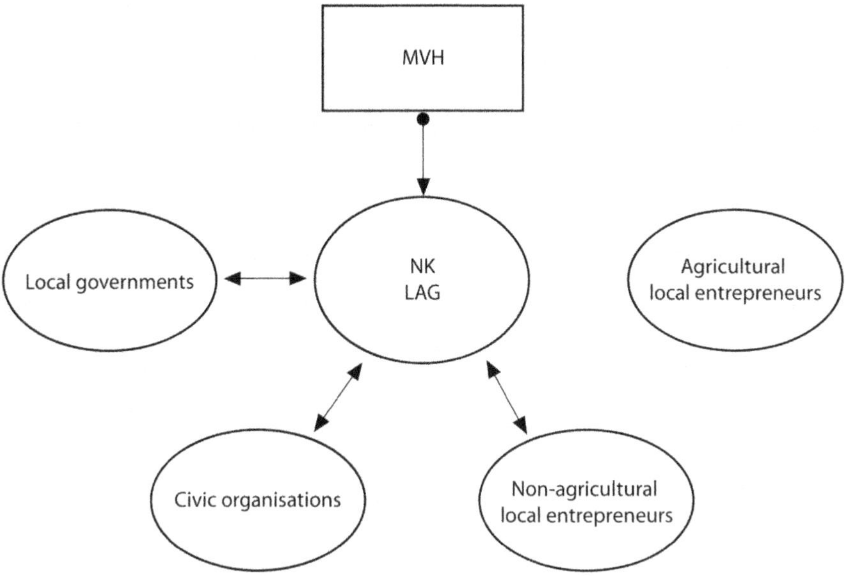

Figure 4.4 Institutional Links of the Nagykunságért LAG

Note: The Nagykunságért LAG demonstrates the limits of actors' involvement in local development. Actors are characterised by their strong knowledge production and reproduction capacity, and their intensive use of different knowledge sets. Having goals, knowledge, the capacity to act, and special skills regarding administrative mechanisms, form the basis of involvement and participation, and determine who gains and who loses in the process (Csurgó et al. 2008).

Decision-making, conflict resolution and knowledge use

The Nagykunságért LAG supported only 33 projects until the end of 2010. They aimed to support financially bigger projects, but did not want to diminish their funding, which amounted to about 342 million HUF (€1,222,000). Because of this, the involvement of the broader local society is very weak.

The decision-making process at a local level can be divided into two phases. The first phase is connected to the LAG personnel. LAG personnel prepare all the documents and tasks before any decision-making. The second phase of local decision-making is connected to the LAG, and especially to the Board, while final decisions are a top-down mechanism from the MVH.

The Nagykunságért LAG does not use explicit conflict-resolving mechanisms, and several conflicts can arise. The main reasons for conflicts are connected to the double decision-making system of the Hungarian LEADER. Local project applicants and project leaders often feel that the LAG personnel have insufficient knowledge and competence, and that they need to be connected directly to the MVH. Project owners evaluate that administrative management is too slow, and they are not aware of the function of the LAG.

Participation in rural-development projects requires knowledge-intensive actors. Many types of local knowledge were used in the case of the Nagykunságért LAG, relating to the level of involvement of local actors. At the same time, the lack of local knowledge causes problems and can lead to failure (Csurgó et al. 2008; Kelemen et al. 2008).

The main knowledge form used by the LAG personnel is that of managerial knowledge. As some interviewees stated, actors with a knowledge of project systems have, locally, a central role in the LEADER system, because of the bureaucratic rules and framework of LEADER. The central role of managerial knowledge has resulted in networking with other LAGs; development actors (e.g. development offices of micro-regions, project-management experts and enterprises, NGOs) are rather less important and visible; the most important partner is the MVH.

Project participation in this region strongly depends on the financial capacity of actors. Actors with capacity have an important involvement in the project, but without capacity they are crowded out and are less able to gather resources (Kelemen et al. 2008; Shucksmith 2000).

The main purposes of the Development Strategy are connected to rural life and tourism, and to the development of non-agricultural SMEs. Most of the supported projects are connected to community building. Civic organisations and local government were active in community-building projects. The second most popular project aim was the development of SMEs. Most of the applicants who implemented enterprise-development projects came from Mezőtúr. The Nagykunságért LAG uses several forms of participation, but most local actors are not involved in the LAG.

The Powers of the Project Class in Local Rural Development – the Case of the Dél-Balaton LAG

Formation and composition of the LAG

The case of the Dél-Balaton LAG demonstrates the emergence of local project-class power in the local-development system. As Kovách and Kucerova (2006 and 2009) have highlighted, experts, designers, European and national administrative staff, holders of intellectual capital and representatives of the civil sector, along with other actors, occupy new social and project class positions. Local project-class members have taken dominant positions in the Dél-Balaton LAG, which has led to several conflicts and the restructuring of power networks and relations.

Actors with intellectual and knowledge capital represent their political, financial and class interests with great potential in a projectified development system, as many scholars have argued in the rural sociology literature on the LEADER system (e.g. Halfacree et al. 2002).

The Dél-Balaton LEADER region is one of the biggest LEADER areas in Hungary, with 33 settlements covering two and a half micro-regions. The centre of the LAG is in Buzsák, but the LAG office is in Fonyód. The legal form of the LAG is a non-profit public limited company. This is the only LAG in Hungary to have chosen this form; most LAGs work as a non-profit enterprise (limited company) or civic association. Several informants emphasised that the form of a public limited company offers transparency and a well-developed structure.

The LAG received 1.5 billion HUF for supporting local projects, and some extra funds for supporting underprivileged settlements; 20 per cent of the total funds can be used for the operational costs of LAG personnel.

There are 165 members in the LAG, all of whom are shareholders in the public limited company (33 local governments, 5 local minority governments, 70 civic organisations and 57 SMEs); 23 per cent of the LAG are public bodies (local governments and minority governments). This complies with the national rule which states that only 30 per cent of LAG members can be local-governmental bodies.

The most important actors are the LAG staff, who take forward the LAG's own issues. The president of the LAG is a local politician, experienced in local development. The General Assembly of the LAG is participated in by all LAG members. Other organisational bodies inside the LAG are: a governing body with nine members; and a board of supervision with eight members elected by the LAG. Only half of the LAG members are active in different LAG organisations and project-application processes. Each organisational body includes all kinds of LAG members.

The most active applicants of the local LEADER are the civic organisations. Community-building through the organisation of local events and tourism development is the most frequent project goal. The activity of local economic actors is very weak. Cooperation of actors is based on regional links and networks. The LAG staff hold the most power, possessing the knowledge, and organising most of the LAG issues. Local governments have still more power in the LAG because of their status and position in the region and inside the settlements, with institutional links to the LAG office and its staff. Only civic organisations have a better position in the LAG, with informal relationships to the LAG staff. Informants emphasised that not all members of LAG appear as project applicants.

Involvement of actors and institutional relationships

The case of the Dél-Balaton LAG follows the theory of Csurgó et al. (2008) concerning the power of the project system over actors. Lack of capacity, cooperation or suitable knowledge leads to the exclusion of actors (Shucksmith 2000). Inclusion depends on suitable capacity, and usable forms of knowledge. Local governments can participate in the LAG as members if they buy shares from the LAG public limited company (Figure 4.5). In the case of the Dél-Balaton LAG all (33) local governments bought shares. Many informants emphasise that, because of political interests of the president of the LAG local governments with a politically left or not-strong right interest have a better position in the LAG. Several local enterprises (SMEs) are involved in the LAG. The LAG personnel have a weaker connection with SMEs than with civic organisations. Local SMEs are mostly inactive as project applicants because of the current economic crisis. They

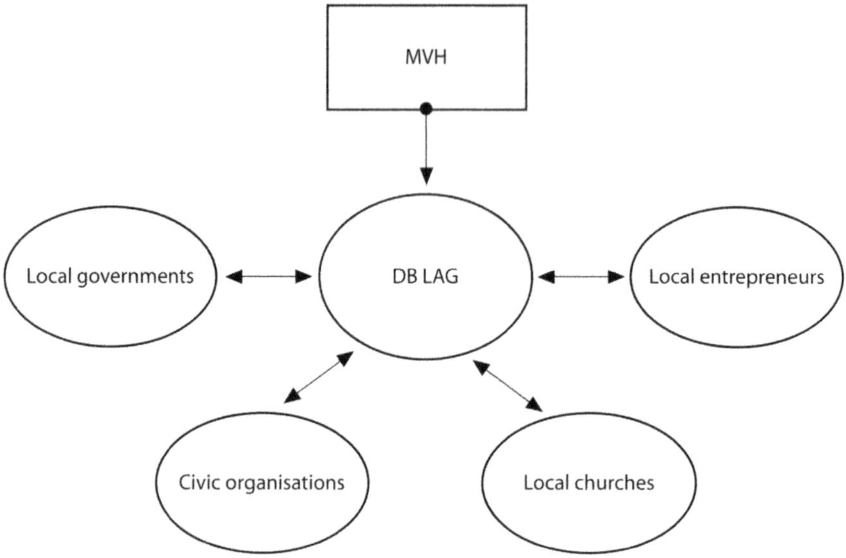

Figure 4.5 Institutional Links of the Dél-Balaton LAG

are not able to finance development projects. The specifics of this region mean that church communities are amongst the most active actors. Almost each local church applies for LEADER funds. They have good informal connections with LAG personnel.

The institutional links of the LAG have two levels. Links with the MVH are bureaucratic and hierarchical; the LAG is under the control of the MVH. At the local level, the LAG has informal connections with different institutions, and the position of LAG in the local development system is hierarchical. Relationships of the LAG with different local actors depend on the connections between them and the level of involvement of actors in the LAG.

Decision-making, conflict resolution and knowledge use

The Dél-Balaton LAG supported only 60 projects until the end of 2010, including 20 European Union financed rural-development projects, 21 LEADER framework projects, and 19 Integrated Community Place framework projects. The total amount of the support was about €5,757,000. The biggest part of the total amount is connected to the Integrated Community Place Framework, and the smallest is concerning to the LEADER. Several LAG members are not active in project applications. According to our results, the exclusion of local actors and LAG members is very strong in the case of the Dél-Balaton region.

LAG personnel use several ways to involve actors, without any success. They organise forums in each settlement to inform local citizens about LEADER and other local development opportunities. They participate in local events, such as

Village Days, presenting the LEADER Community as an important actor in the local regional society. Nevertheless, the involvement of actors is based on their network position in the local development system. We can find a hierarchical power-knowledge structure dominated by local elites, controlling the development process and power structures as a result of a projectified local development system. The involvement of actors in that kind of projectified local development system is based on the capacity and network of actors. Cooperation and participation is a chance to reach the goals of actors (Csurgó et al., 2008). Some respondents pointed out that political conflicts had arisen between LAG members and LAG personnel. Conflicts occurred in the case of tourism projects. There were some projects which the LAG had to refuse because of a lack of money. These project applicants feel they have a disadvantaged position because of different interests, and a weak network position.

Civic organisations are very active applicants, as are LAG members. An other type of member expressed the view that local civic organisations received too much support, while there were not enough funds for economic development because there were too few motivated applicants among SMEs. They complained that LAG personnel have not done enough to motivate local economic actors.

The Dél-Balaton LAG does not explicitly use conflict-resolution mechanisms. In this LEADER region, tourism is the most important economic sector. The general regulation of LEADER for 2007–2013 stated that actors in the mass tourism areas do not apply for rural tourism development funds. Several settlements from the region, mostly from the Fonyód micro-region are excluded from such LEADER funds. This form of exclusion causes several conflicts and disadvantages where the tourism-based settlements could stimulate the local economy, as well as tourism for the whole region.

The regulated system of local rural development resulted in involved actors who need special knowledge to put forward a development project. Managers and bureaucrats have appeared and have played important roles in the project process. Most of the supported projects are connected to community building. Civic organisations and local government were active in community-building projects. They organised several local events. The most successful was situated in Kisberény, where there is a large Roma minority population. The result of the project is that this local minority became involved in local community life and events. The second successful project aim was the preservation of local heritage buildings. Many churches in the region received renovation funds.

All the successful projects and project leaders have strong connections with LAG personnel, including local project-class members. Those projects supported in the Dél-Balaton LAG region are strongly controlled by LAG personnel, as a part of a local project-class. Network positions provide the basis for involvement and participation, and determine who gains and who loses in the process of local rural development.

Summary of Case-Study Results

The results of the case studies are presented in Table 4.2 according to the themes and issues used for the focus of the analysis.

Table 4.2 Case Study Results

Formation and composition of the Local Action Groups			
	HAVER LAG	**Nagykunságért LAG**	**Dél-Balaton LAG**
Number of LAG members	large	low	large
Position of local governments in LAG	non-central	central	non-central
Main LAG actors	non-agricultural entrepreneurs	local governments and non-agricultural enterprises	LAG personnel and civic organisations
Involvement of actors and institutional relationships			
Role of MVH	bureaucratic power	bureaucratic power	bureaucratic power
Role of LAG Personnel	motivational	bureaucratic	central power
Decision-making, conflict resolution and knowledge use			
Project participation of locals	wide	narrow	very narrow
Characteristics of supported projects	tourism development	community building and development of SMEs	community building
Knowledge use	lay and expert knowledge	expert knowledge	administrative knowledge

An important characteristic of LEADER in Hungary is the strong control of MVH as a national bureaucratic authority, as demonstrated by the case studies. The system is over-bureaucratised, which contradicts with the bottom-up principle of LEADER. However, LEADER is important in the local development process, but LEADER regions differ in the operation and implementation, which results in differences between the local democracy and the involvement of local society in the local development process.

The number of LAG members, the role of local governments in LAG, and the types of main LAG actors determine the involvement and participation of local actors, and also determine the operation of the LEADER system from project application to decision-making at the local level. The case of the HAVER LAG presents a relatively high level of local democracy. Large numbers of LAG members, the weak power of local governments, and the strong position of local entrepreneurs and civic organisation in the LAG prove the relatively wide integration of local stakeholders in the local LEADER development system. The case of the Nagykunságért LAG, with fewer actors and a stronger position of local governments, along with the high importance of knowledge and the capacity of actors in the involvement of local development, demonstrate the way in which the position of the elite in local societies is supported and strengthened by the LEADER system. Finally, the case of Dél-Balaton LAG has presented the highly bureaucratised ways of LEADER at a local level, where the LAG personnel have had a strong administrative role and the LAG works as an administrative body of local development at the local level. It is characterised by the weak participation of local actors and the domination of administrative knowledge and power in the local development system.

Conclusions

The case studies give much evidence that LEADER has great potential in the local development process; however, many informants criticise the political management of local LEADER. As they explain, LEADER practice does not meet European Union principles and it is not a real bottom-up development system. LAGs do not have real power over decisions. Local actors were unable to localise development; all decisions are controlled by the Agricultural and Rural Development Agency, MVH, and national bureaucratic authorities. According to such criticism, the system is over-bureaucratised, which contradicts with the bottom-up principle. Several local actors feel that this system is much more characterised by top-down principles than bottom-up ones. LEADER and the local action groups are instruments which channel development funds to local actors; however, the three case studies presented demonstrate that the challenge of profit-earning through projects may devalue the bottom-up character of LEADER.

The case studies in Hungary do not paint an overly optimistic picture of decision-making in Local Action Groups, which is seen as a more aggregative democratic process than the case of integrative democracy. Andersson and Kovách argue that 'the model of integrative democracy may seem to be the right point of departure if one aims at analysing LAGs/LEADER, making use of core democratic theory' (2010, 16). After analysing the case studies, one may ask if LEADER practice is a sign of democracy in Hungary or not.

Comparing case studies, it is obvious that many elements of LEADER activity in Hungary are formally analogous with each other, for example, the legal form of the LAGs is that of non-profit enterprise. Through the participation and involvement of civic, governmental and economic institutions and actors, formal democratic decision-making comes to fruition. The LAG staff plays a decisive role, and their managerial knowledge is of great importance. From another side, we can say that locals do not have enough independence. The most powerful actor is MVH, the national authority. The result is that the LAG does not have real power; the LEADER system in Hungary is characterised by elitism. It demonstrates the illusion of bottom-up development.

The main differences between LAGs, and also LEADER regions in Hungary, are connected to the participation process and the outcomes of the local LEADER-type development. Our cases demonstrate different types of participation in local development. The HAVER LAG has adopted several forms of participation, from forums to project collection. The LAG gives financial assistance to several smaller projects. The aim is to support a variety of actors from all kinds of settlements. The Nagykunságért LAG also uses several forms of participation, but most local actors are not involved in the LAG; instead, they support bigger projects. The Dél-Balaton LAG does not succeed, despite many attempts, in integrating their members into LEADER. Local actors and applicants are under the control of powerful actors, who are especially connected to the LAG personnel.

In the case of the HAVER LAG, most of the supported projects are connected to rural tourism and rural heritage. In the region of the Nagykunságért LAG, most of the supported projects are connected to community building. Civic organisations and local government have been active in community-building projects. Another popular project aim has been the development of SMEs. Renovating community buildings and protecting local heritage buildings were the most popular project aims in the case of the Dél-Balaton LAG. Supported applicants also organised local events. Many churches received support for renovation.

LEADER-type development can be seen in Hungary as the first step for local democracy. According to our data, the system is over-bureaucratised and over-controlled. The predominance of general rules results in the exclusion of several important rural actors from the system (e.g. agricultural entrepreneurs cannot get support from LEADER). From the other side, according to the Hungarian case studies, the experiences are not only negative. LEADER has been able to contribute to a rising dawn of a bottom up system: learning and networking; the emergence of cooperation; sharing influence amongst stake-holders; an innovative project class which replaces monopolistic domination and the power of local economic and political elites; strengthening local identity; revitalising local culture. All this may provide a basis for the advent of local democracy.

References

Andersson, K. and Kovách, I., 2010. Lagging behind or leader in local democracy?: An assessment of LEADER-type development projects as a tool for democratic integration in the contested countryside. SSKH Reports and Discussion Paper; Nr 3/2010. Helsinki: Swedish School of Social Sciences, University of Helsinki, p. 34.

Aunapuu-Lents, V., 2012. The LEADER approach and the accumulation of power in the case of Estonia. *Administrative Culture*, 14(1), pp. 125–44

Bruckmeier, K., 2000. LEADER in Germany and the Discourse of Autonomous Regional Development, *Sociologia Ruralis*, 40 (2), pp. 219–27.

Buszt L., Vedres B., 2013. Local developmental agency from without. *Theory and Society*, 42, Issue 1, pp 1–23.

Csite, A., 2005. *Reménykeltők*. Budapest: Századvég.

Csurgó, B., Kovách, I. and Kučerová, E., 2008. Knowledge, Power and Sustainability in Contemporary Rural Europe. *Sociologia Ruralis*, 48(3), July 2008, pp. 292–312.

Csurgó, B., Kovách, I. Légmán, A. and Megyesi, B., 2010. Energy demand, Governance and Infrastructure in Hajdú Bihar County: The Hungarian Case. In: Kovách, I. and Gotts, N., eds, 2010. *Climate Change and Local Governance: Alternative approaches to influencing household energy consumption: a comparative study of five European* regions, Budapest: MTA Politikatudományi Intézet, pp. 185–224.

Dragos, D. and Neamtu, N., 2007. Reforming local public administration in Romania: Trends and obstacles. *International Review of Administrative Sciences*, 73(4), pp. 629–48.

Fazekas, Zs. and Nemes, G., 2005. Kísérleti Leader-jellegű program Magyarországon. In: *AVOP Leader+ készségek elsajátítása*, Promei Kht, Faluműhely alapítvány, SZRVA, pp. 457–82.

Furmankiewicz, M., 2009. Enchancing endogenous development in rural areas: the implementation of the LEADER Programme in Poland [Online]. Available at: <www.asecu.gr/files/RomaniaProceedings/24.pdf>. (Accessed 13.12.2014.)

Halamska, M. and Maurel, M.C., 2006. *Les acteurs locaux à l'*épreuve *du modèle européen LEADER: France – Hongrie – Pologne*, Prague: CEFRES; Varsovie: IRWiR PAN, p. 206.

Halfacree, K., Kovách,I. and Woodward, R., eds, 2002. *Leadership, Local Power and Rural Restructuring in Contemporary Europe*. Aldershot: Ashgate.

Harghita County Council, ed., 2005. *Area of Action and Target Group Report. Leonardo daVinci*. Miercurea Ciuc, Romania.

High, C. and Nemes G., 2007. Social learning in LEADER: Exogenous, endogenous and hybrid evaluation in rural development. *Sociologia Ruralis*, 47(2), pp. 103–19.

Hudeckova, H. and Lostak, M., 2008. Agriculture and farming related activities: their actors and position in LEADER approach. *Agricultural Economics* [*Zemědělská ekonomika*], 54, č. 6, s., pp. 245–62.

———, 2010. Preliminary impacts of the LEADER+ approach in the Czech Republic. *Agricultural Economics* [*Zemědělská ekonomika*], roč. 56, č. 6, s., pp. 249–65.

Kelemen, E., Megyesi, B. and Nagy Kalamász, I., 2008. Knowledge Dynamics and Sustainability in Rural Livelihood Strategies: Two Case Studies from Hungary, *Sociologia Ruralis*, 48(3), pp. 257–73.

Knickel, K and Renting, H., 2000. Methodological and Conceptual Issues in the Study of Multifunctionality and Rural Development, *Sociologia Ruralis*, 40(4), pp. 512–28.

Kovách, I. and Kucerová, E., 2006. The Project Class in Central Europe: The Czech and Hungarian Cases. *Sociologia Ruralis*, 46(1).

———, 2009. The Social Context of Project Proliferation – The Rise of a Project Class, *Journal of Environmental Policy & Planning*. 11(3), March, pp. 203–20 (16).

Kovách, I. and Kristóf, L., 2009. The Role of Intermediate Actors in Transmitting Rural Goods and Services in Rural Areas Under Urban Pressure. *Journal of Environmental Policy & Planning*, 11(1), March, pp. 45–60 (16).

Lee, J., Árnason, A., Nightingale, A. and Shucksmith, M., 2005. Networking: Social Capital and Identities in European Rural Development. *Sociologia Ruralis*, 45(4), pp. 269–84.

Mathieu, P. and Pascal, M., 2010. Actors and partnership in the LAGs: a new civic space? A Czech case study. In Plésiat_Marty - Lag as a New Civic Space [Online]. Available at: <http://www.docstoc.com/docs/73627380/Pl%C3%A9siat_-Marty---Lag-as-a-New-Civic-Space>. (Accessed 14.12.2014.)

Maurel, M.C., 2008. Local Development Stakeholders and the European Model: Learning the LEADER Approach in the New Member States. *Sociologický* Časopis / *Czech Sociological* Review, 44, pp. 511–29.

———, 2009. Local actors facing environmental issues: lessons for a bottom-up approach [Online]. Available at: <http://hal.archivesouvertes.fr/docs/00/50/30/99/PDF/8_Maurel_2009_Local_actors_facing_environmental_issues.pdf>. (Accessed 2.12.2014)

Nemes, G., 2005. A Magyarországi vidékfejlesztés eddigi tapasztalatai, a Leader+ intézkedés tartalmi elemei és alkalmazásának szabályai. In: *AVOP Leader+ készségek elsajátítása*, Promei Kht, Faluműhely Alapítvány, SZRVA, pp. 331–54.

Osti, G., 2000. LEADER and Partnerships: the Case of Italy. *Sociologia Ruralis*, 40(2), pp. 172–81.

Ray, C., 2000. Editorial. The EU LEADER program: rural development laboratory. *Sociologia Ruralis*, 2/2000.

———, 2001. Territorial co-operation between rural areas: elements of a political economy of EU rural development. *Sociologa Ruralis*, 41(3), pp. 279–95.

Sharman, J.C., 2003. Agrarian Politics in Eastern Europe in the Shadow of EU Accession. *European Union Politics*, 4, pp. 447–71.

Shortall, S., 2004. Social or Economic Goals, Civic Inclusion or Exclusion? An Analysis of Rural Development Theory and Practice. *Sociologia Ruralis*, 44(1), pp. 109–23.

Shucksmith, M., 2000. Endogenous Development, Social Capital and Social Inclusion: perspectives from leader in the UK. *Sociologia Ruralis*, 40(2), pp. 208–18.

Shucksmith, M., and Chapman, P., 1998. Rural Development and Social Exclusion. *Sociologia Ruralis*, 38(2), pp. 225–42.

Sjöblom, S., 2006. Towards a projectified public sector – project proliferation as a phenomenon. In: Sjöblom, S., Andersson, K., Eklund, E. and Godenhjelm, S., eds. *Project Proliferation and Governance*, Helsinki: Helsinki University Press, pp. 9–33.

Tovey, H., 1998. Rural actors, food and the post-modern transition. In: Granberg, L. and Kovach, I., eds. *Actors in the Changing European Countryside*, Budapest: HAS Institute for Political Sciences.

ÚMVP, 2009. *Jelentés. Az Új Magyarország Vidékfejlesztési Program végrehajtásának 2008.* évi *előrehaladásáról* [Online]. Available at: <http://www.umvp.eu/files/umvp_eves_jelentes_2008.pdf>. (Accessed 13.12.2014.)

Varga, E., 2009. Non-profit organizations in Hungarian rural development: a LEADER + example in a Southern Transdanubian region. *European Countryside*, pp. 99–104.

World Bank, ed., 2008. *Romania. Municipal Finance Policy Note. Report no. 38357-RO*, 3 June 2008. World Bank, Poverty Reduction and Economic Management Unit.

Chapter 5
The Democratic Capabilities of and Rhetoric on LEADER LAGs in the EU – The Danish Case

Annette Aagaard Thuesen

Introduction

This chapter assesses the contributions of partnership organising in Danish LEADER local action groups (LAGs) to democracy from the viewpoint of aggregative and integrative democratic theory. The most important players in the LEADER programmes over the years have been the LAG board members. Thus, the core of the LEADER method is the establishment of a LAG partnership consisting of representatives from the public, private and voluntary sectors. The LAG partnership holds decision-making power, creates a development plan and provides project grants to project holders; it can also initiate its own projects. In addition to the partnership concept, the LEADER method consists of six other key concepts: bottom-up approaches, area-based strategies, cross-sector/integrated approaches, innovation, cooperation, and networking. The LEADER method for rural development has been the backbone of what has come to be known as the New Rural Paradigm (OECD 2006), in which the focus is on area-based/territorial rural development rather than sector-based rural development.

Through the LEADER programmes, the EU has supported the transfer of decision-making authority to LAGs. The EU's support can be viewed as an expression of the need to introduce new players into the rural development scene. States cannot secure rural development alone, and the initiative has been transferred to other players, including actors from the private and voluntary sectors. These new players are important for the proper implementation of the decisions made by central authorities. The role of partnerships is central because partnerships function as governance networks developed from a steering philosophy based on political power functioning as a form of empowerment through enabling, rather than coercing, people to participate (Bang 2005).

The structure of LEADER in the EU has changed from a Community Initiative during LEADER I (1991–1993), LEADER II (1994–1999) and LEADER+ (2000–2006) to being mainstreamed into the rural development program (RDP) and fisheries program (FP) of 2007–2013 (CEC 2006). As a Community Initiative, LEADER functioned as a laboratory for investigating new solutions for rural development. As part of the RDP and FP in Denmark, LEADER currently utilises a methodological approach to rural and fisheries development in areas of the RDP and FP in which more sector-specific aims related to agriculture, forestry and environmental conservation are also pursued (CEC 2005; CEC 2006a).

The chapter proceeds in the following manner: after this short introduction, the second section presents the theoretical framework by which the democratic capabilities of Danish LAGs is analysed followed by a presentation of the methods in the third section. In the fourth section, first the EU rhetoric regarding LEADER is analysed followed by an analysis of the empirical data on LEADER in Denmark during the LEADER+ and RDP and FP 2007–2013 periods. Finally, the last section offers concluding remarks.

Theoretical Background

New Institutionalism

This chapter views LAGs as rather formalised governance networks, or institutions, in a wider system of governance, in which they participate in the allocation of the resources of a society. The new institutionalist approach provides a suitable framework by which to understand these processes.

Normative institutionalism (founded by March and Olsen) assigns a central role to norms and values within organisations and believes that individuals are influenced at all times by their full range of institutional attachments. Institutions mould their participants and supply systems of meaning (Peters 1999, 26). Thus, institutions are not necessarily a formal structure, because a collection of norms, rules, understandings and routines plays just as important an institutional role (Peters 1999, 28; March and Olsen 1989). The central term of normative institutionalism is the 'logic of appropriateness', which influences individual behaviour (March and Olsen 1989, 160–62). Peters posits the following regarding organisational norms: 'If an institution is effective in influencing the behavior of its members, those members will think more about whether an action conforms to the norms of the organization than about what the consequences will be for him- or herself' (Peters 1999, 29). The 'logic of consequentiality' is thus less important in the behaviour of organisational members, and the basis of institutions is therefore normative rather than coercive (Peters 1999, 41).

Examining LAGs from the perspective of normative institutionalism provides a good understanding of the formation of the identities of LAG board members through their participation on a LAG board. Sørensen and Torfing (2007a, 37) write: 'The identities of the network actors are shaped and reshaped through socialization, education, reflection, action and interaction which is conditioned by the logic of appropriateness that has been developed within the governance network'. The logic of appropriateness will differ between different LAG boards.

The logic of consequentiality will increasingly fade into the background as the willingness of board members to contribute to the work on the board increases.

Rational choice institutionalism, however, is based on the classical assumption that individuals make utility-maximising decisions. Thus, it contrasts with March and Olsen's approach. Rational choice institutionalism attaches importance to institutions as mechanisms for channelling and constraining individual behaviour (Smed 1998, 121) because institutions are viewed as a precondition for the interaction of utility-maximising individuals. In this respect, individuals *can* make rational decisions to be constrained by memberships in institutions (Peters 1999, 44). Without the presence of institutions (often in the form of rules), rational individual actions might generate collective irrationality. The rational choice approach is concerned with the design of institutions based on the assumption that the behavioural element of individual maximisation will lead individuals to shirk their responsibilities (Ostrom 1990, 27; Peters 1999, 45). Institutions can easily be created with different costs and benefits (incentive structures) that are not the consequences of historical and norm-based processes. Instead, game-like situations occur, as expressed by the governability theoretician Scharpf (1997).

Assessing LAGs from a rational choice institutionalist perspective places emphasis on rational actors depending on each other to achieve common policy goals. If each actor cannot achieve a goal individually and the benefits of cooperation are higher than the costs, it is rational to join networks. This approach also draws attention to the importance of establishing rules, incentive structures and games as political instruments to encourage trust between the parties. Sørensen and Torfing (2007b, 34) write: 'The preferred games are those that emphasize the positive interdependence of the actors and increase the costs of non-cooperation'. Rational choice institutionalism emphasises the fact that each LAG board member has his/her individual reasons for joining the LAG board.

Aggregative and Integrative Democracy

March and Olsen's (1989) two approaches to democracy – the aggregative and the integrative approaches – can be viewed as the continuation of rational choice institutionalism and normative institutionalism (Bogason 2004, 3). The theory outlined by March and Olsen is based on the belief that the organisation of political life in the form of institutions is important. It goes beyond the rational choice perspective of individual decision-making because March and Olsen take the normative institutionalist approach:

> Political actors are driven by institutional duties and roles as well as, or instead of, by calculated self-interest; politics is organized around the construction and interpretation of meaning as well as, or instead of, the making of choices; routines, rules, and forms evolve through history-dependent processes that do not reliably and quickly reach unique equilibria; the institutions of politics are not simple echoes of social forces; and the polity is something different from,

> or more than, an arena for competition among rival interests. (March and Olsen 1989, 159)

Political institutions as instruments of democracy will be evaluated differently from the perspectives of aggregative and integrative democracy. March and Olsen acknowledge the integrative democratic stance in their assumption that the identities and capabilities of individuals are closely related to their membership and position within a community (March and Olsen 1989, 161). They write:

> In a broad sense, we have been urging that a perspective of politics as organized around the interaction of a collection of individual actors or events may be supplemented with (or replaced by) a perspective that sees the polity as a community of rules, norms, and institutions. (March and Olsen, 171)

The aggregative democratic version is based on the traditional institutions of representative democracy, wherein representatives are elected to address the interests of individual voters. Between elections, the voters are not to participate in the discussions among the representatives. Instead, the representatives' performance will be judged on Election Day. Society is viewed as a collection of individuals with individual and atomised preferences that must be protected. Negative liberty, as asserted by Berlin (1958), is therefore central to the aggregative democratic standpoint.

The integrative democratic version is built on the idea of deliberation towards an identification of the common good. From this stance, the role of representatives is to promote deliberation, which makes citizen participation crucial. Through participation and deliberation in the civil society and in the political institutions, individuals will be educated and transformed into citizens (Bogason 2004, 5–6; Sørensen and Torfing 2007b, 234–5). Contrary to the aggregative version, democracy is not viewed as a power struggle between different individuals or factions but rather as a forum for deliberations regarding the common good. Liberty is thus defined positively (Berlin 1958). March and Olsen's integrative position is evident in the following quote:

> Even in situations in which there is ex ante disagreement about values, there are processes of public discussion and private thought that arrive at better ex post social solutions than does bargaining, exchange, or coalition formation in the service of prior preferences. (March and Olsen 1989, 127)

Examples of Aggregative and Integrative Democratic Theoreticians

A democratic theoretician who takes the aggregative stance as a point of departure is Schumpeter (1994). Schumpeter is a proponent of elite democracy. He finds that it is unrealistic and inappropriate for citizens to directly participate, as democracy should instead consist of elites competing for votes. When defining democracy,

Schumpeter states, 'the democratic method is that institutional arrangement for arriving at political decisions in which individuals acquire the power to decide by means of a competitive struggle for the people's vote' (Schumpeter 1994, 269). According to this definition, what is left for voters to decide is the question of accepting or rejecting the chosen representatives who, according to Schumpeter, should be professional politicians. Democracy is a method for the selection of leaders, and the role of citizens is reduced. Schumpeter believes that he has created a realistic model because citizens do not ultimately achieve any real influence, nor should they have any influence because they do not have the expertise and capability to deal with political problems.

A democratic theoretician who departs from the integrative position is Pateman (1970). She seeks to broaden the scope of citizens' contributions to decision-making and thereby expand the role of citizens beyond casting votes. Pateman is concerned with the role of participation in modern democracies. She asserts that individuals and their institutions cannot be considered in isolation from each other (Pateman 1970, 42). Participation is necessary both at the local level and in industry to obtain a democratic polity. The main functions of participation are educational, including the psychological aspect and the acquisition of practical democratic skills. She characterises the participatory democratic model as 'one where maximum input (participation) is required and where output includes not just policies (decisions) but also the development of the social and political capacities of each individual, so that there is "feedback" from output to input' (Pateman 1970, 43). According to Pateman, participation has an integrative effect and therefore aids in the public acceptance of collective decisions.

In the Danish context, the aggregative and integrative versions of democracy have been exemplified in the writings of Ross (1967), who defines democracy as a type of allocative structure similar to representative democracy, and Koch (1945), who defines democracy as a way of life. Ross and Koch differ in their view of the capacity of citizens to participate in politics. Ross does not find that ordinary people have these capacities for the following reasons:

> The immediate public decision is unreliable, not only because people often due to ignorance, instability and short-sightedness will fail in their assessment of what serves it well, but also because often the majority will be determined by ominous special interests at the expense of a reasonable account of the minority. (Ross 1967, 220)

By contrast, Koch finds that people are capable of learning to participate in democracy, as he asserts that democracy:

> Is a mindset, a lifestyle that cannot be acquired 'til you live through it in a very narrow private life in relation to family and neighbors, and then outwardly in relation to larger circles, in the relationship to fellow countrymen, and finally in relation to other nations. (Koch 1945, 12–13)

According to Ross, the task of the people in a democracy is therefore not to govern directly but rather to choose representatives in the form of elite bureaucrats and leaders who will be better able to govern (Ross 1967, 220). Koch is more in favour of direct citizen participation in political decision-making because this approach serves an important educational purpose.

The aggregative stance and the integrative stance, the stance of Schumpeter and Pateman and the position of Ross and Koch, respectively, are summarised below in Table 5.1 to demonstrate the adversarial horizontal relationships between the three vertically congruent positions.

Table 5.1 Aggregative and Integrative Democracy (top level), Democracy According to Schumpeter and Pateman (middle level) and According to Ross and Koch (bottom level), Respectively

Aggregative	Integrative
'Political systems are similar to economic systems built around competitive markets and prices'.	'Human rights are seen as inviolate and inalienable symbols of the integration of a political culture, as well as of the identities and commitments of citizens'.
'The democratic method is that institutional arrangement for arriving at political decisions in which individuals acquire the power to decide by means of a competitive struggle for the people's vote'.	'The participatory model as one where maximum input (participation) is required and where output includes not just policies (decisions) but also the development of the social and political capacities of each individual, so that there is "feedback" from output to input'.
'People need leaders. The thought about the individual's self-governance and responsibility should be connected to the thought about leadership in trust'. 'Idea about leadership in acknowledgement of and trust in other's larger insight and capacities'.	'Democracy cannot be confined in a formula. It is not a system or doctrine. It is a way of life. It is the conversation (the dialogue) and the mutual understanding and respect that are the nature of democracy'.

Methods

Before presenting and discussing the empirical data, a brief methodological overview is needed. The empirical portion of this chapter is divided into three sub-sections based on differing data collection methods. The first sub-section presents a brief analysis of the overall EU LEADER rhetoric, which forms some of the foundation upon which LEADER in Denmark has been built. The second sub-section is a review of the overall institutional design of the governance system of LAGs in the Danish LEADER+ 2000–2006 and RDP/FP 2007–2013 context. Here, the empirical data were obtained from document studies of policy documents as well as knowledge collected through research projects. The third sub-section presents qualitative interview data related to LEADER's contribution to democracy. This sub-section is based on a recent qualitative analysis (Thuesen and Nielsen 2012) of LAG board members and coordinators' view on the benefit of the LEADER method, which included democracy as a theme. Focus group interviews were conducted in five LAGs in the period from October 4 to November 1, 2011, and 4–7 persons were interviewed at each focus group meeting. The data were transcribed and subsequently analysed by use of concept based coding categories in accordance with the theoretical framework used in this chapter.

LEADER: The EU Architecture and the Design and Function in Denmark

Assessment of the EU LEADER Rhetoric

According to March and Olsen, concrete institutions can be judged from the perspective of aggregative and integrative democracy because institutions embrace specific conceptions regarding the goal of political processes. Bearing this fact in mind, it is possible to assess the extent to which the LEADER approach in the formal presentation of the program at the EU level builds upon the aggregative stance or the integrative stance. Such an assessment can be made by examining the fact sheet 'The LEADER approach – A basic guide' (CEC 2006b). In this publication, the European Commission mentions the integrative aspects of LEADER numerous times, as demonstrated in Table 5.2

Table 5.2 Assessment of the Aggregative and Integrative Aspects of LEADER at the EU Level

Aggregative or integrative judgment	Quotations from 'The LEADER approach – A basic guide' (CEC 2006b)
Integrative	'The LEADER approach can provide an opportunity for them to take the initiative and to participate actively, in rural development programmes in their local area'.
Integrative	'By encouraging local participation in the drawing up and implementation of sustainable development strategies, the LEADER approach may prove to be a precious resource for future rural policy'.
Integrative	'It uses a holistic approach to address rural problems'.

Aggregative or integrative judgment	Quotations from 'The LEADER approach – A basic guide' (CEC 2006b)
Integrative	'LEADER encourages socioeconomic players to work together, to produce goods and services that generate maximum added value in their local area'.
Integrative	'An area-based approach takes a small, homogenous, socially cohesive territory, often characterised by common traditions, a local identity, a sense of belonging or common needs and expectations, as the target area for policy implementation'.
Integrative	'Local actors participate in decision-making about the strategy and in the selection of the priorities to be pursued in their local area'.
Integrative/ aggregative	'The involvement of local actors includes the population at large, economic and social interest groups and representative public and private institutions. Capacity building is an essential component of the bottom-up approach'.
Integrative	'LAGs … strengthen the dialogue and cooperation between different rural actors, who often have little experience in working together, by reducing potential conflict and facilitating negotiated solutions through consultation and discussion'.
Integrative	'Putting these principles into practice means real people designing local strategies and participating in activities'.

Table 5.2 shows how the active participation of citizens is clearly encouraged at the EU level. The fact sheet mentions that the LEADER programme uses a holistic approach to rural development – holism being the opposite of atomistic – which stands in opposition to the aggregative stance. Moreover, capacity building is highlighted as a central component of the LEADER approach, along with strengthened dialogue and cooperation between different socioeconomic players. Aggregation of interests is mentioned only in the context of LAG partnerships that must be representative of the existing local interest groups drawn from different socioeconomic sectors in the area (CEC 2006b). The overall LEADER rhetoric thus appears to be integrative.

Assessment of the Institutional Design of Danish LAGs

Concerning the institutional design of LAGs in Denmark, the situation is more diverse. In contrast to the 12 Danish LAGs in the LEADER+ period of 2000–2006, 57 LAGs have been established as part of the current RDP and FP 2007–2013 in Denmark. This statistic means that more than 700 persons are active on the LAG boards (Thuesen 2010; Thuesen and Sørensen 2008). In total, 39, 12 and 6 LAGs work strictly within the RDP, within both the RDP and FP and strictly within the FP, respectively. Together with an increase in the number of LAGs, a reform of the organisational arrangement has occurred from LEADER+ until RDP and FP 2007–2013. During the LEADER+ period, the members of the LAGs were appointed rather than elected. Thus, it could be difficult for an outsider to gain entrance to or influence decision-making processes in the LAG partnerships. During the LEADER+ period, LAG board members were called LAG members because in most places, there were no additional members of the LAG aside from the board itself. During the programming period 2007–2013, the formal institutional design established by the Ministry prescribes that it is obligatory for Danish LAGs to be organised as associations with free and open membership. People who live in the LAG area and are more than 15 years can become members of the LAG association and participate in the annual election of the board. The board members, who must be at least 18 years old, are elected for two-year terms at annual general assembly meetings held in the spring. The board members are to represent four groups: 1) local citizens, 2) local enterprises and trade organisations, 3) local nature, environment, culture, citizen, and leisure associations, and 4) public authorities (DFIA 2007a; DFIA 2007b). The situation prior to 2007 was not democratic in either the aggregative or the integrative manner. Input legitimacy was simply too low due to the appointment structure.

Table 5.3 Overview of the Danish LAG Organisation from 2000 to 2006 and from 2007 to 2013

	LEADER+ 2000–2006	RDP and FP 2007–2013
Number of LAGs	12	57
Number of board members	Approximately 184	Approximately 700
Entry standards to LAG boards	Appointment	Election of people over 18 years old, elected for two-year terms at annual general assembly meetings
Additional LAG members	No	Yes, free and open membership for people who live in the area and are more than 15 years old

The establishment of a higher number of LAGs has given more citizens the opportunity to influence local decision-making. The fact that LEADER covers nearly the entire country (with the exception of several large cities) has resulted in a much greater awareness of the work LAGs perform within the wider rural population, as well as greater media attention.

To strengthen democracy in relation to LAGs, clear access procedures are necessary, such that all citizens have the opportunity to influence decision-making. Such was not the case in the Danish LAGs during the LEADER+ period. The situation has changed with the introduction of LAGs as membership-based associations. Nonetheless, in relation to participation in the decision-making of LAGs, LEADER in Denmark is still a 'democracy of the active'. Only those individuals who join the association and attend the general assembly meetings can influence the composition of the board. As demonstrated by Thuesen (2010), Danish LAG board members in the RDP and FP 2007–2013 are extremely well educated, which means that they could be deemed sub-elites (Etzioni-Halevy 1993). They have the expertise and capability to address political problems, contrary to what Schumpeter (1994) and Ross (1967) believe is true for the general population. Elections improve accountability structures and formal representation. This aspect points towards an increased concern regarding the aggregative aspects of democracy. From a ministerial viewpoint, such concerns might be founded upon the expectation of increased public awareness due to the greater amount of money distributed though the LAGs in the period of 2007–2013. Drawing upon Schumpeter's argument, the competitive struggle for the people's vote has been introduced and enhanced with the increased media attention and greater public awareness. However, participation patterns continue to prove uneven. According to Pateman and Koch, these patterns are problematic due to the organisations' inability to fulfil the educational function of participation (Pateman 1970; Koch 1945). Authority has been transferred to an organisation that citizens are to actively research through newspaper articles, advertisements and websites. From the perspective of the aggregative democratic theories, this process is not necessarily problematic because the central focus is on the election of representatives who are best equipped to make decisions. From the perspective of the integrative democratic theorists, however, this process is problematic due to the lack of involvement of new citizens. Thus, it does not fulfil the educative or capacity-building functions of LAGs. For LAGs to be organised as associations also requires the involvement of more people through the membership structure. This arrangement can be viewed as an opening towards an integrative democratic position through various types of involvement. In this respect, the Danish context reveals LAGs to be useful democratic supplements to existing political structures. Overall, it can be said that elections have improved the accountability structures of Danish LAGs, pointing towards the aggregate version of democracy, and the new membership model and the greater awareness of the programme among the media and the general populace hold the potential for the development of the integrative democratic stance, as well.

Assessment of the Work Accomplished Through the LAG Boards

The establishment of LAGs and the discussions within the LAG boards are described as unique and special by many LAG board members. To borrow Pateman's concept, the boards engage in a process where there is feedback from output to input. One board member states:

> What I think is great about the whole thing here, the years I have been in the LAG system [is] that it brings people together across dividing lines. That is, you bring business people together with people like us … together with some politicians, and they are therefore forced … you have to sit down around a table and discuss it, and say what we want with the development. I think you have no other forum in Denmark, where … it can be done. Otherwise the professional organisations sit and say, 'No, this is my money, I'll make sure of that'. And then the nature organisations say that here it should just be nature all together. And then there are some politicians who say, 'Ah, we cannot even afford it'. Here, they are obliged to talk to everyone, and that is what I think is the most amazing thing about the LEADER approach. You bring some people together who would normally never get to meet in that way. And even that one can develop things and use some money too. I think that's the genius of this system. (Interview 1, 2011)

One could say that the fact that both the Ministry and the EU provide money if the local partners create a development plan together helps to emphasise the interdependence between the players and increases the costs of not cooperating, as described in the section on rational choice institutionalism. The local players may 'lose face' locally, and they can also lose money if they fail to reach an agreement on the development plan. The Ministry and the EU have set up a game structure as a political instrument to build trust between the socio-economic actors. However, this approach only appears to prevail during the initial phase of the board's work. A board member for another LAG explains how agreeing on the strategy led to some serious discussions, during which the aggregation of interests dominated initially but was replaced by an understanding of the importance of achieving a common outcome through dialogue and collaboration rather than through competition:

> It was exciting and it was good that we got further into it. Because there was a time when we thought, now we all go home, and now nothing more will happen here … where we all went in and gave what we had to it … we could see the point in continuing with it because we had a common interest in seeing that it got up and running. (Interview 2, 2011)

A representative of the same LAG Board elaborates on the topic: 'It is also a strength that … there are different people in such a group that … Yes, it could well be that one could sit and hoe his own potatoes … but it's no use in a community [like the LAG]'. Again, the norms appear to be well established and more important than individual self-interests. Thus, there is a clear expression of collaboration rather than competition in the concrete work of the LAG boards. A LAG board member also describes the importance of discussions between board members when making decisions as follows:

> I think it's exciting, it's always exciting to read the projects before you've talked with anyone about them, and come only with one's own prejudices, or try to identify them or what to even think about anything, and then to evaluate them by what is being said. Therefore, I think that a discussion on the board is incredibly important because we have such a different approach to the projects. (Interview 3, 2011)

Another board member discusses the same subject as well as what motivates him to continue working with the LAG as follows:

> After I have participated on the LAG board meetings, I think an additional motivating factor is the diversity of the people who are present. And even though we do not always have an equal amount of time to read all the projects, it does not really matter because we have each understood our part of it. And all together, I think that when we talk about the projects, we always discuss them in depth. And I also think that this quorum is an inspiring thing. (Interview 3, 2011)

Thus, there appears to be a solid consensus about the previously mentioned March and Olsen quote (March and Olsen 1989, 127), which states that even in situations in which there are differences among individuals' values initially, there are processes of public dialogue and private reflection that arrive at better social solutions in the end than would bargaining or coalition formation in the service of goals established in advance. There is also a positive recognition that the LAG board is a forum that brings different people together. One might ask whether the discussions would become more competitive if there was a greater sense of scarcity for funding.

When asked about the benefit of the LEADER method, a board member directly expresses that it is a project that supports democracy as follows:

> It's a way … both what is happening in the LAG … the group here; it is a motley mix of people who come from all sorts of places. I think there's an interesting discussion, when it is not a professional selection in the way you read the applications, and the way applications are assessed through a variety of lenses. And I think that it matters in the long run also that people know … I think part of the LAG method's success is that people know that there are local people who sit at the table here and … [that] there is a dialogue about what is happening … In that way, I think it is a project that supports democracy. … The LAG is of course much better known than many other funding pools. It obviously has something to do with the time … that it has been around for so long. But it also has something to do with, I think, the local anchorage … the local decision-making. (Interview 3, 2011)

Another person describes democracy as beneficial in the following manner:

> We may well say that another approach to strengthening democracy, it's exactly also that. [On] the board … we largely come from everywhere, that is, there are not many where you can say that they come from [the bigger towns]. They come from all the small communities of which we have 21 or 22 in the municipality. (Interview 3, 2011)

Some of the common ground among board members is established by virtue of board members being local people from local communities and decisions being made locally by a broad circle of board members. A LAG board member refers to this local decision-making and coaching as follows:

> I also believe that the LAG's success and hopefully future success is exactly related to that local decision. But [it is] also something about [the fact] that you can grab a local coordinator, so it is not [an organisation] that sits in Odense or Copenhagen or Brussels. So you can get to talk about, and you can get out and see, and you may be advised: is this something we have an opportunity for at all? So it's not all done by mail or by phone. We can pull project initiators in if we have doubts about whether it is justified that we should give grants. (Interview 3, 2011)

Overall, it is the logic of appropriateness that internally prevails on the LAG boards rather than the logic of consequentiality. Individual preferences for joining the LAG board appear to become less important over time compared to the development of a territorially based consensus and an integrative version of democracy.

Conclusion

In conclusion, the associational institutional design, including elections of board members, has created room for processes of integrative democracy that focus on involving more citizens in democratic discussions and promoting more deliberations and openness around the LAGs in general. The elections place LAG partnerships somewhere between a pure network structure and a traditional representative structure. The introduction of elections can thus be viewed as an expression of the move toward a more aggregative democratic approach, in which elections are meant to establish the legitimacy of the LAGs and in which sub-elite competition is viewed as an optimal democratic solution. Regarding the work on the LAG boards, these processes are ongoing. Moreover, as the assessment above demonstrated, this work is primarily based on principles of integrative democracy, in which communities of rules and norms develop over time.

Finally, I posit that experiments at the LAG level and at the project level can contribute to the further democratisation of the LEADER initiative in Denmark. Ideally, these experiments could be accompanied by national guidelines or nationally initiated experiments or games focused on inclusion. They would encourage participatory practices among all LAGs. By initiating broad projects on, for example, village development plans and concrete citizen involvement, LAGs can become important mediators in rural governance between formal authorities (i.e. the EU, state, region and municipality) and rural citizens.

References

Berlin, I., 1958. Two Concepts of Liberty. In: Berlin, I., 1969. *Four Essays on Liberty*. Oxford: Oxford University Press.

Bogason, P., 2004. *Institutional Theory and Democracy*, Working paper, 2004(7), Roskilde: Centre for Democratic Network Governance.

CEC, 2005. *Council regulation No. 1698/2005 of 20 September 2005.*

———, 2006a. *Council regulation No. 1198/2006 of 27 July 2006.*

———, 2006b. *The LEADER Approach – A basic guide*. Luxembourg: Office for Official Publications of the European Communities.

DFIA, 2007a. *Vejledning om oprettelse af og tilskud til drift af lokale aktionsgrupper i landdistrikterne* [*Guidance on establishing and subsidising the operation of local action groups in rural areas*]. Danish Ministry of Food, Agriculture and Fisheries / Danish Food Industry Agency.

———, 2007b. *Vejledning om oprettelse af og tilskud til drift af lokale aktionsgrupper i fiskeriområderne* [*Guidance on establishing and subsidising the operation of local action groups in rural areas*]. Danish Ministry of Food, Agriculture and Fisheries / Danish Food Industry Agency.

Etzioni-Halevy, E., 1993. *The elite connection – problems and potential of western democracy*. Cambridge, MA: Polity Press.

Hall, P.A. and Taylor, R.C.R., 1996. Political Science and the Three New Institutionalisms. *Political Studies*, 44, pp. 936–57.

Koch, H., 1945. *Hvad er demokrati?* [*What is democracy?*]. København: Gyldendal.

March, J.G. and Olsen, J.P., 1989. *Rediscovering Institutions. The Organizational Basis of Politics*. New York: The Free Press.

Ostrom, E., 1990. *Governing the Commons. The Evolution of Institutions for Collective Action*. Cambridge: Cambridge University Press.

Pateman, C., 1970. *Participation and Democratic Theory*. Cambridge: Cambridge University Press.

Peters, B.G., 1999. *Institutional Theory in Political Science – The 'New Institutionalism*. London and New York: Continuum.

Ross, A., 1967. *Hvorfor demokrati?* [*Why democracy?*]. København: Nyt Nordisk Forlag Arnold Busck.

Scharpf, F.W., 1997. *Games real actors play – Actor-centered institutionalism in policy research*. Boulder, CO: Westview Press.

Schumpeter, J.A., 1994. *Capitalism, Socialism & Democracy*. London and New York: Routledge.

Smed, J., 1998. Bottom-up og nyinstitutionel teori – Teoriudvikling er nødvendig [Bottom-up and new institutionalist theory – Theory development is needed]. In: Bogason, P. and Sørensen, E., eds. *Samfundsforskning Bottom-up – Teori og metode* [*Social Research Bottom-up – Theory and Method*], Roskilde: Roskilde Universitetsforlag, pp. 119–45.

Sørensen, E. and Torfing, J., 2007a. Theoretical Approaches to Governance Network Dynamics. In: Sørensen, E. and Torfing, J., eds. *Theories of Democratic Network Governance*, Palgrave Macmillan, New York, pp. 25–42.

———, 2007b. Theoretical Approaches to Democratic Network Governance. In: Sørensen, E. and Torfing, J., eds. *Theories of Democratic Network Governance*, New York: Palgrave Macmillan, pp. 233–46.

Thuesen, A.Aa., 2010. Is LEADER Elitist or Inclusive? Composition of Danish LAG Boards in the 2007–2013 Rural Development and Fisheries Programmes. *Sociologia Ruralis*, 50(1), pp. 31–45.

Thuesen, A.Aa. and Nielsen, N.C., 2012. *Merværdien af LEADER* [*The added value of LEADER*]. IFUL Report 16/2012, Esbjerg: SDU.

Thuesen, A.Aa. and Sørensen, J.F.L., 2008. *Lokale aktionsgrupper i landdistrikter og fiskeriområder 2008 – sammensætning, aktiviteter og samarbejde i opstartsperioden* [*Local Action Groups in rural and fisheries areas 2008 – composition, activities and cooperation in the start-up phase*]. IFUL Report 9/2008, Esbjerg: SDU.

Interviews

Interview 1-3, 2011. Interviews of LAG board members and coordinators. See details from Thuesen A.Aa. and Nielsen, N.C. 2012.

Chapter 6

A Political Perspective on LEADER in Finland – Democracy and the Problem of 'Troublemakers'

Marko Nousiainen

Introduction

LEADER in Finland

Finland is one of the most rural countries in Europe. According to the Organisation for Economic Co-operation and Development's (OECD) definitions of rural areas, Finland ranks fifth in terms of the share of territory covered by predominantly rural regions (89 per cent), and second both in terms of population that it hosts (53 per cent) and GDP produced in these regions (45 per cent) (OECD 2008, 32). However, at least since the late 1960s, rural areas have faced serious challenges caused by urbanisation and agricultural modernisation. Both the rural livelihoods and the amount of population have been diminishing. In response to the problems, the Finnish government began to set up specific rural policy initiatives in the late 1980s. This development led to the emergence of today's rural policy institutions, which are geared to enhancing the wellbeing of the population living in the rural areas. One of the most prominent of these is the LEADER approach started after the country's membership in the European Union in 1995 (Hyyryläinen 2007; European Communities 1988). The purpose of this chapter is to critically examine the Finnish LEADER and the prospects of democracy it offers[1]. In the following pages, I will evaluate the prospects and problems of LEADER as a democratic institution focusing mostly on perspectives of consensus and conflict.

The principles of bottom-up and local empowerment have been taken seriously in the implementation of the LEADER approach in Finland. At least in the composition of Local Action Groups (LAGs) this kind of purpose is visible. LAGs in Finland are third sector actors, registered development associations in which anyone can become a member. The executive body of a LAG is the board and it is elected by the members of the association in the Annual General Meeting (AGM). At least in theory, any citizen living in the area of a LEADER LAG can, thus, become a member of the development association and even get elected to sit

1 See also Rosenschöld and Löyhkö in Chapter 2 of this volume.

in the board. The board has a tripartite structure: one-third of the board members represents public authorities (mostly municipalities and their officials), one-third represents local NGOs and one-third represents 'ordinary' citizens, people who do not belong to any of the other two quotas. Consequently, we can think that the empowerment of larger societal groups has been 'built in' in the composition of the Finnish LAGs. Most of the decision makers in LAGs are rural residents and many are even laymen in policy matters.

The LAG boards make decisions in matters belonging to their 'sphere of authority'. For example, decisions about which projects will be funded are a central part of the work that the LAG boards carry out. The powers to grant funding to individual projects have been divided between the LAGs and the regional government authorities (ELY centres) so that LAGs make decision concerning the usefulness of individual projects and the ELY centres make decision about their legality. Thus a project application needs to get approval in both systems before it is accepted as a LEADER project. LAGs also have permanent offices and employed staff of two to four persons. Each LAG prepares a local development plan for its operating area and the plans are implemented with public funding.

Largely because of its localness, LEADER action in Finland – as abroad as well – has been evaluated as a novel form of local democracy. Writers such as Pylkkänen (2004), Pylkkänen and Hyyryläinen (2004), Kull (2008), Wade and Rinne (2008) and Ramos and Mar Delgado (2003) have discussed LEADER as an experiment of democracy. It is sometimes said that LEADER could be seen as an initiative in participatory democracy that renews or supplements the representative political system in Finland (e.g. Karhio 2000; Wade and Rinne 2008). Thus, LEADER is not seen merely as a rural development scheme but also as a novel way to understand political action and participation. Yet, the plausibility of this notion has not been properly scrutinised on the basis of empirical material focusing on the 'grass-roots' level participation. Instead of declaring it as a novel form democracy, we should examine LEADER and its relation to political action. This is the objective of this chapter.

Methods

The focus of this chapter is the Finnish LEADER on the grass-roots level. The study is based on my doctoral theses (Nousiainen 2011), in which I wanted to get first hand insight about how the participants in LAGs actually consider their participation. This is why ethnographic methods, interviews and participant observation were chosen. I consider the grass root level and every day understanding to be important when we evaluate the political or democratic capacities of LEADER. As well as abstract ideas of democratic theory, every-day practices and grass-roots level understanding are important tests of the democratic qualities of LEADER.

During the field work period (2007–2008), I carried out 18 semi-structured interviews among the active LAG participants in four different LAGs. Most of the actors I interviewed live and act in the Eastern parts of the country. Nine of these were men and nine were women. Five of the interviewees were LAG employees, managers or other paid staff, and 13 interviewees were members of the boards. I adopted an open-ended strategy in the interviews since I wanted to let the actors speak out their own views, experiences and opinions concerning their participation in LEADER. In the interviews a set of questions was presented that dealt with the successes and difficulties faced by the LAG, the interviewees' personal motivation to participate and the relation of politics and LEADER action, to mention a few. In many cases it was not the questions that were leading the discussion. If the interviewees wanted to discuss some other subject, it was acceptable as long as the interviewee's experiences or opinions concerning LEADER were the general topic. All the interviews were recorded and transcribed verbatim.

When interpreting the material, I am interested in the discursive construction of LEADER action. The object of study in discourse analysis is how the socially constructed objects of reality are created and how they are sustained (Phillips and Hardy 2002, 6). This kind of method gives a special role for the use of language. Language is not seen only to represent reality, but language use is seen to construct and form our perception of it. Language use has a great impact on how we think and act, even if it does not fully determine our subjectivity. In this analysis, I seek accounts describing the Finnish LEADER and read ways of how the interviewees are giving meaning to their own participation. In these interview discussions, the participants defined LEADER as a specific field of human action and also defined themselves as actors. In the analysis, I read different sequences, narratives, comments and remarks in which the interviewed actors give substance to their personal perspectives concerning LEADER. Especially, I am interested in accounts that can be related to democratic and political qualities of LEADER: for example meanings that deal with understanding consensus and conflict among the participants of the LAG boards. I do not think that these meanings could be right or wrong. Rather, they all are interesting sections of discursive construction which have a great impact on what kind of mental schemes guide the action and every-day practises of rural development. On top of interview speech, written material, published government documents, studies and evaluations are possible sources through which we can study what kind of human action the participation of local people in LAGs should be seen.

In addition to discursive construction of LEADER, I am also interested in the resulting practices. During my field work period, I conducted also some participant observation. (see Nousiainen 2011). I took part in the every-day work of one Local Action Group in Eastern Finland and participated in, for example, board meetings, AGMs and ordinary work of the LAG office. I wrote a field diary of the time in the LAG. In this chapter, I also make some references to my experiences in the field. Since an anonymous treatment was promised to all those who took part in my research project as informants, I will not give any detailed descriptions of the cases, the LAGs, or name any of the interviewed persons.

Theory: Democracy and Politics

In this chapter, the conceptions of consensus and conflict have a central position. These are also a central theme in political theory, including the theory of democracy. March and Olsen (1989) divide democratic ideas into aggregative and integrative theories. In the aggregative theories, democracy means the aggregation or gathering of different views and preferences into a system of collective decision-making. Conception of conflict is emphasised in the aggregative theories: democratic decision-making means choosing one of conflicting policies through political competition or bargaining. The integrative theories, on the other hand, emphasise deliberative action. According to the deliberative views of democracy the point is to form – through public discussion – such a collective decision that could be accepted by all (Setälä 2003, 131). Rather than aggregating different preferences, deliberative democracy is about changing them. Especially Habermasian ideas of communicative ethics and action can be seen as an important inspiration of deliberative democratic theory (Setälä 2003, 132). Habermas is often cited also in the Marsh and Olsen's (1989; 1995) views of integrative democracy.

March and Olsen (1995) see democracy mostly as a deliberative and as a transformative category. Democracy means changing the actors rather than aggregating their views or preferences in the system of collective decision-making. Democracy is about educating citizens, fostering democratic identities and capabilities. While the citizen is seen as a political actor, integrative democracy defines her also as an object whose democratic character should change. The goal of this change is to produce a democratic culture in which conflict is civilised, debate is reasoned and that is geared to producing common good. In this process, March and Olsen (1995, 34–6) define a base of shared meanings, rules and cultures as vital even if they accept that such assumptions are often problematic. A romantic view of a homogeneous culture may be dangerous. Despite of this, certain view of community, a life shared by people and some kind of consensus over common good is an important premise of their political thinking. There must be a base of shared meanings (that may mean shared identities or common language or practices) if deliberative processes are going to succeed at all. As Hobson (2009, 178) has noted, there often is a strong circular dimension in the ideas of deliberative democracy since they expect certain cultural requirements to be met before deliberation can begin. Deliberation, thus, necessitates the kind of common understanding it should be producing. However, it must also be noted that the point of March and Olsen (1995) is not to eliminate conflicts but to form such a political culture in which rational and civilised deliberative action is possible (March and Olsen 1995, 60).

Even if we acknowledge that certain kinds of shared beliefs are prerequisites for any notion of democracy,[2] the urge for a shared culture and consensus can also be seen as dangerous especially in contemporary post-modern and fragmented

2 For example that conflicts should be solved with words instead of weapons.

societies. Pluralism, at least, can be seen as problematic for the deliberative notion of democracy. The more there is pluralism in the views, interests, values or lifestyles in society the less likely the deliberating assembly is to find a common ground from which rational agreements can be reached. The problem of pluralism is visible also in the hostility with which the deliberative democracy theorists discuss ideological and rhetorical forms of reasoning. March and Olsen (1989, 133), for example, claim that deliberative democracy theory requires a shared rationality that is not corrupted by ideology or rhetoric. As also Urbinati (2006, 119–20) has shown, theorists of deliberative democracy see any ideological or rhetorical forms of communication as perverse forms of reasoning. Therefore they seem to deem the traditional parliamentary mode of action and debate as an insufficiently rational form of communication to be seen as proper deliberation.

Also the notion of common good is important to March and Olsen (1995, 49) who even define democracy as a means to pursuit it (March and Olsen 1995, 49). It is assumed that when the conditions for rational deliberation are perfect such an ideal will follow. The idea of common good is as well prone for practical problems. If a common ground is not reached by the means of un-coerced deliberation, it may become lucrative for actors to adopt some less rational means. And one can imagine how this kind of situation might turn the rational processes of deliberation into their perversions, such as 'group think' as Bill Cooke (2001, 108–20) has noted.

In spite of seeing democracy only as a system of producing integration, common good and consensus, we should take the ideas of aggregation and conflict more seriously. The conflict view of democratic politics allows us to examine competing notions of common good and different forms of reasoning as an integral part of a democratic culture. According to the Finnish political scientist Kari Palonen's (1993, 2003a) Weberian notion of politics as politically motivated action, politics means performative actions in the struggle over a share of power. Palonen (2003a, 184) defines four common uses for the word 'politics': politicking, politicisation, policy and polity. This polit-vocabulary forms a framework under which different uses of 'politics' can be analysed. Politicking and politicisation are the main aspects in understanding politics as action. Politicking is defined as action in the struggle over power and politicisation enlarges the sphere of politics and thus enables performative political actions. Politics is a sphere of contingency where competition and the creative use of rhetoric freely reign. Acting politically also means accepting conflictual positions and struggle, but also forming and destroying coalitions. Pluralism is not a problem for Weberian notion of politics; rather it is its precondition (Palonen 2003a, 172–6).

Policy, can be defined as normative-teleological line of action or a project or a plan that coordinates political actions in order to fulfil some preferred goals. Policy, like integrative democracy, requires some kind of a shared conception of the goal, the common good and the means to achieve it. Polity, then, means a political sphere, political community or an arena in which politicking may take place. It can be a geographic area or an entity of things already assumed as political, as contingent. Therefore, policy and polity are concepts that seek to delimit politicking and politicisation (Palonen 2003a, 172–6.). Yet, even polity and policy should also be seen as political concepts. Both are results of some previous political act or debate (Palonen 2003b, 470).

I am suggesting in this chapter, that in addition to integrative notions of democracy we should consider adopting the perspective of politics in the research of LEADER approach. Different uses of the word politics give us tools to analyse what kind of political action the discursive production of LEADER is promoting among the participants and what kind of a polity it is creating. This is important if we seek, as I do, to evaluate the democratic potential of the LEADER action. The problem of pluralism in the theory of deliberative democracy, at least, requires a critical scrutiny of the local level conceptions concerning LEADER. I think that the crucial question for deliberative democracy is what happens when the context of deliberation is not an ideal one. In real world the situations of reasoned discussion – as all other situations as well – are always less than perfect. Therefore the idealist aspects of deliberative democracy may rather lead us astray than give us useful insights for empirical analysis.

LEADER on Local Level: the Ideal of Consensus and the Problem of 'Troublemakers'

According to the calculations published by Pylkkänen (2004, 84–5),[3] it is clear that Local Action Groups are not such mass movements that involve large proportions of their area's population in direct participation. An average association serving as a LAG has some 150 individual members. While a LAG's area may consist of several municipalities and have many dozens of thousands of inhabitants, one cannot claim that some members form a large portion of the population. If we look at the participation in the LAGs' annual meetings, this tendency is even more striking: the average participation in 2003 was only two dozens of participants and even the best figure of the early 2000s was 69.[4] The number of people participating in the making of the local LEADER development plans is significantly higher.

According to Pylkkänen (2004, 76–7), about 6,000 people have participated in the making of the programmes of LEADER+ period (2000–2006). Nevertheless, this forms only 0.7 per cent of the total population of the LEADER areas. Yet, it is also unclear what kind of influence this participation has had on the actual plans. In general, these figures indicate that it is not plausible to think that the LAGs form a credible forum in which large masses of people can take-part in public decision-making or indeed an institution of representative democracy. However, the more

3 Pylkkänen's data is from the LEADER+ programming period. Nevertheless, there is no good reason to assume that significant changes would have occurred during the current period.

4 The smallest number I have heard of is two and the biggest number is fifty.

interesting question is whether they constitute a forum in which political debate, different world views and values or interests of different societal groups are engaged in deliberation. This requires the inner logic and discursive construction of LEADER to be studied.

According to the interview material, it seems clear that a discourse emphasising consensus is widely shared among LEADER activists. This discourse also carries normative connotations, thus constructing a conception of LAG board's ideal inner dynamics. In other words, in the discursive construction of the Finnish LEADER action there is a strong tradition suggesting that disputes or conflicts should not be tolerated within a well-functioning Local Action Group. This ideal is visible in many materials scrutinised in this study, for example interview speech concerning several different interview topics as well as texts published by the Ministry of Agriculture and Forestry. When asked about the relations between LAG board participants, some interviewees were eager to define their group as harmonious and unanimous in the execution of its mission. Also the narratives describing the problems faced by the LAGs often envisage this idea. The discourse of consensual relations among participants can also be noticed in the interview discussions concerning political action.

In a couple of interviews the harmonious relations among the participants in a LAG board were defined as a crucial factor to the success of LEADER action locally. Here, consensus does not necessarily mean that participants should agree on all the matters discussed. Rather, it means a shared mental state, a common consciousness about the goals of LEADER action and the values behind them. This reminds of a shared rationality that is seen important in the theory of deliberative democracy (March and Olsen 1995, 34–6). Yet, this kind of shared consciousness was sometimes defined far from trouble-free in the interview material.

The role of the ideal of consensus can be illustrated through examination of two vivid narratives in which the consensus discourse was described in terms of trouble faced by the LAGs. These narratives were presented by actors of different LAGs and in both of these sequences the basic story was the same: conflicts among the participants and especially among the core persons in the LAG may be very harmful. The first of these interviewed actors answered a question about the problems faced by the LAG with a story of a difficult year four years ago (Interview 6). According to him there had been a difficult person in the LAG, a chairman of the board, with whom the employed staff did not get along. As a consequence, the trust between actors was shattered. The board was divided into smaller groups that held their own meetings and, thus, the group was not able to communicate properly. According to the interviewed LAG manager, the lack of trust made operation very difficult and it could have even meant the end of the LAG – at least in the worst case scenario. In the narrative, the LAG seemed incapable to cope in front of this sort of difficulty.

In his story, the interviewed actor (Interview 6) gave no other reason for the conflict than the personal character of the ex-chairman. The conflict seemed to have no substance. This was typical of all such interview sequences where group dynamics were defined as problematic. At the same time, the groups that were defined as ideal were never described to be a consequence of the personal traits of a participating person. This may indicate that consensual relations among LAG actors are seen as an ideal notion, that needs to be protected with the identification of 'the guilty persons' when faced with difficulties. It gives the participants a possibility to think that a shared consensus is real – as long as difficult persons are excluded.

Other vivid narrative about the perils of conflict was told by another LAG manager in her interview (Interview 2). She also described incidents that had happened years ago. According to her story, she was working as an employee in a LAG that's inner dynamics was frustrating. Board members were very critical towards most of the project applications. They were masters in finding downsides in the applications even when the employees would have been more optimistic. This was a difficult situation to the employees who felt that they had to argue against the board. The reason for the negative group dynamics in the board was similar to the narrative discussed above. There was one person who was the source of the negative culture, 'a troublemaker', and another board member who supported him. Accordingly, the group dynamics was defined a fragile thing that even one person not willing to act correctly may destroy. Therefore she saw it important to focus attention to it.

The story told by the interviewee had a happy ending (Interview 2). When the 'troublemaker' and his supporter left the LAG board, the group reached a more consensual working culture. The participants considered themselves a single group that shared the same goals and values. In the story, the best way to solve problems of group dynamics was getting rid of those participants that were a hindrance to the ideal relations between members of the group. In this interview, also, conflicts among participants were defined as very destructive to the efficient functioning of a LAG: the group seemed incapable of solving differences through discussions. Again, no other reason for the conflict was presented than the bad behaviour of the participant. In the light of this experience it is possible to understand why the character of the participating individuals may be seen as of crucial importance.[5]

The interview material suggests that the ideas of reasoned deliberation and consensus have a big part in the discursive construction of LEADER action. Yet, the material also shows that the context for deliberation may often be far from perfect. Unlike the theory of deliberative democracy suggests, it is not the 'preferences' but the personalities of some of the participants that cause trouble – and this is viewed as a big problem. The two narratives of 'troublemakers' show that, instead of deliberation, exclusion is seen as the plausible solution to this problem. Unfortunately, I had no possibility to interview the other parties of these conflicts and therefore I only have observations that are probably one-sided. Even

5 The problem of 'troublemakers' is reminiscent of 'bad behaviour' noticed by Newman and Clarke (2009, 177). Their observation was that community, civil society or voluntary participants may take their 'empowerment' too seriously or literally.

if the interviewed actors defined personal characters as the source of the problems, some substantial differences may also have been behind these disputes. However, both possibilities lead to quite pessimistic consequences for the idea of LAG as a deliberative forum. LAGs seem incapable to solve disputes – either personal or substantial – by the means of democratic deliberation.

The normative discourse of consensus is visible also in other parts of my interview material. I asked the interviewees, active participants in Local Action Groups, to comment one of the topics that I found interesting: the relation of politics and LEADER action.[6] The resulting speech was mostly in concert with the discourse of ideal consensus discussed above. Nevertheless, the interviewees seemed to be well aware of the conflictual nature of political action within the institutions of representative democracy. The LEADER action was often defined in relation to the representative political bodies, most commonly municipal councils or parties. One interviewee, for example, described participation in a LAG as a form of civic activism that is consensual in nature whereas politics in municipal council is full of conflicts and disputes (Interview 8). When municipalities make hard decisions on how to use the money collected from the area, LEADER action groups have an easier job deciding about money that is not directly collected from the pockets of the local tax payers. This was described as a fortunate thing since it allowed the action groups to avoid issues that would have been probable sources of conflict.

Especially the use of the term 'party' in the interview discussions shows an interesting aspect of the important role of consensus in the discursive construction of LEADER action. Even if participation of different type of actors is one of the core rationales of the Finnish LEADER, none of the interviewees told me that the participation of political parties should be acceptable. If nearly twenty informants share the same view unaware of each other's answers, it is probable that the discourse is widely shared among the actors. My interpretation is that the concept of party symbolised a sort of conflictual action and ethos that the interviewees wanted to keep away from their group. Party as a symbol of political conflict is nearly the opposite to the ideal of consensus that most of the interviewees held in so high regard.

In general, many of the interviewees defined the role of political conflict as a core difference between the municipal politics and LEADER action. As one board member told me, LEADER should be geared to efficient execution of rural development goals instead of 'irrelevant quarrelling' – as she defined municipal politics (Interview 9). A LAG should have a common goal and the participants should work together to accomplish it. This view was shared even by many of those interviewees who took a more positive stance on politics (e.g. Interview 1; Interview 2; Interview 14). LEADER was defined as a rural development policy with certain goals which the interviewee saw important. In other words, it was

6 In my interview questions I did not give any definition for political action since I wanted to make the interviewees to deploy their own understanding of the concept.

defined as a normative and teleological form of action. A number of policy goals were mentioned: equality between regions, social objectives and care for the environment – to name a few (e.g. Interview 10). Yet, none of the interviewees told me that a LAG could be an arena in which the debate over such goals should take place. Rather, these goals were always defined as undisputed. It seems clear then, that the interviewees defined LEADER action as participation in a well framed policy system that has its objectives established. The participation can thus be interpreted as voluntary participation in the realisation of a pre-given policy rather than discussion of what this policy should consist of. If we accept Palonen's (1993; 2003a) notion of politics, LEADER is for the interviewees more like a means to accomplish some political goals, a normative and teleological line of action, a policy, instead of a space of politics as a struggle over a share of power.

This interpretation of the interviewees' speech constructs an ideal type of what perfect LEADER action should look like. As the horror stories about the 'troublemakers' demonstrate, the every-day LEADER action may not be as consensual as the actors would want it to be. In addition, some interviewees also presented narratives about the corruptive influence of politics, which they had experienced during their careers as LAG activists. It was said, for example, that LAG employees' credibility will surely suffer if they take part in municipal elections (Interview 2) and some board members had tried to benefit politically of their position in the LAG board (Interview 16). In these interview comments, political action was seen as an awkward matter that should not be a part of LEADER action even if it sometimes is. The politically active participant could also be seen as a kind of 'troublemaker' who causes difficulties for the smooth functioning of LEADER in the local level.

The problem of 'troublemakers' is taken seriously on behalf of the administration, too. A number of responses to the problem have been developed. For instance, some traits in the composition of the Finnish LAGs make it possible to choose the participating individuals more carefully. All the LAGs in Finland are third sector organisations, registered associations, and therefore their members are legally free to elect anyone they please to represent them in the board. In practice, this freedom is a bit more restricted. LAG boards have tripartite structure that orders all members to represent one of the three quotas: public sector, NGOs and 'ordinary' rural inhabitants.[7] Also the representation of different geographical areas or municipalities is seen as important. This combination makes the work of the AGMs electing the boards a fairly complex task: election of one 'ordinary' rural inhabitant necessitates electing respectively two other board members in the other quotas.[8] Therefore – according to participants of a board meeting that I observed

7 This is stated in the contract all development associations make with the government.

8 According to some of the interviews, the ordinary rural inhabitants are, surprisingly, the most difficult quota to fill (Interviews 5 and 6). Ordinary inhabitants can also be 'fabricated' by asking otherwise suitable persons to give up their other commitments and thus free themselves to act as ordinary residents in LAG boards (Interview 2).

during my field work period – it was not seen plausible to expect that AGMs would be able to fulfil the demands of tripartition without a careful preparation. At least each new board member must be asked their agreement beforehand. It seems to me that the boards or the LAG offices are the active parties in this preparation work instead of the common members of the association who may only have a faint idea of the regulations directing the execution of the LEADER approach. The tripartition gives, in fact, a possibility to choose the core participants in advance. When the board members and employees are contemplating possible future members of the LAG board they will surely consider the participants' other qualities in addition to their position in the tripartition. Therefore, the tripartite structure of the LAG boards can be used to exclude certain kind of persons from participation, as well as it can be used to serve a 'democratic' end.

Another trait that allows the LEADER offices to have an impact on the participating individuals is the rule that limits the board members' 'term-of-office' to six years. An idealist observer, as I was, would understand that the six-year-rule serves a democratic function allowing larger groups to participate. But the interviews strongly suggested that this was not the case. Rather, it could be defined as a matter of efficiency as it made it obligatory for LAGs to renew their working culture (Interviews 11, 12 and 15). In addition, I heard also another explanation for the rule which I found peculiar. The six-year-rule was initiated because it allowed the LAGs to get rid of such board members that were not suitable for the task, the troublemakers for instance. Especially such board members that were seen as difficult, uncooperative or unskillful could be fired from the group on the basis of the rule. 'The good ones' can be taken back to the board after a year's rest. Two interviewees – who were experienced and seemed to know well the ideas of the national level of LEADER administration – gave this explanation for the rule (Interviews 2 and 14). In addition to interview discussions, this view is presented also in a study published by Uusitalo (2009, 175), the long-term chairman of the Finnish Rural Policy Committee. The will to select cooperative persons to act as LAG board members was, therefore, shared also on the national level of the Finnish rural development administration.

The combined practices of the six-years-rule and tripartition make it possible to administer participation in the LAG boards. Even if most of the participants are laymen living in the LAG's area, the practices make it possible to select members so that no 'troublemakers' could get in, in the first place. Therefore these national traits in the composition of LAGs have a role in helping the ideal of consensus to become a reality. This means that the aspirations in the grass root and the national level LEADER administration may have a little to do with enhancing the ideas of local politics or pluralism. Rather on the contrary. They help to make the participating group more uniform in order to establish a shared base of meanings and values so that deliberation would be efficient. The undisturbed functioning of every-day LEADER administration seems to be the foremost objective. A smooth and efficient functioning is an understandable objective when LEADER is seen as a policy that has its goals and its means to accomplish them.

The strong discourse of consensus is evident also in some of the texts published by the Ministry of Agriculture. A self-evaluation guide for the LAGs (Vehmasto, Vuorio and Lahtinen 2003) gives a detailed technique for how to teach uniform attitudes, values and goals for LEADER actors. Pylkkänen (2008, 227) has described this evaluation guide as a practice of knowledge and power that strives to change the thinking of the participants. As well as the regulations concerning the composition of the LAGs, also evaluation can be seen as a practice that works to enhance a culture of consensus within LAGs.

According to this examination, the views of LAGs as functioning systems of democracy – integrative or aggregative – seem dubious. LEADER does not obviously involve large masses of people or their 'preferences' in the decision-making process. Also the discursive construction of LEADER action seems to highlight certain shortcomings of the theory of deliberative democracy. Rather than suggesting dialogue between opposing views, or other civilised and democratic qualities, the strong discourse of consensus constructs a culture helping participants to think and act according to uniform mentalities. The resulting conception of participation does not encourage political debate, but rather works to achieve an efficient execution of the rural policy objectives. Moreover, the personal character of the participant is emphasised: he or she should share the same underlying values as others or simply be willing to conform. As Hobson (2009, 178) noted, deliberative democracy seems to require the kind of consensus it should be producing. Thus also the Finnish LEADER LAGs seem incapable of dealing with conflicts unless certain kinds of persons are excluded from participation.

Conclusions

The strong discourse of consensus observed in the study has consequences when thinking of LEADER as a new form of political participation and action. The discourse constructs LEADER action as a specific form of human endeavour that is not easily compared to other phenomena of political reality. The LEADER discourse constructs a new kind of political subjectivity that differs from more traditional ways to understand politics. This may not only represent a possibility to our understanding of democracy, but it may also be a threat to it.

Participation in the LAGs can be viewed as political, since the participants often define it as a way to advance certain political objectives that they see important. The participants also exercise certain power in the LAGs and make important decisions in the field of rural development. The strong discourse of consensus evident among participants suggests, nevertheless, that rather than local forums of political debates, LAGs should be considered as instruments of rural policy. With a policy different political acts are gathered together to support certain teleological line of action (Palonen1993; 2003a). LAGs look like organisations whose role is circumscribed to the execution of a pre-given rural development strategy and which have local resources based on voluntary participation at their disposal.

Strong consensus among participants is probably useful in the execution of shared strategies and therefore the urge towards consensus is understandable. If we accept Palonen's (1993; 2003a) Weberian view of politics as conflictual action, struggle over a share of power, the apolitical nature of the Finnish LEADER discourse is striking.

The policy view on Finnish LEADER seems to be consistent in various materials examined in this study, including the local voluntary participants' interview discussions. LEADER being a central part of the national and EU's rural policy implementation, this result should not be seen as a surprise. Yet, the discussion of democracy, so typical to many studies addressing LEADER in Finland and abroad, raises a concern over democracy. As a result, democracy may be reduced to action and participation in governance systems where political debate and criticism are replaced by an administrative ethos of accepting strict pre-defined frames for political imagination. In this kind of process, the theories of integrative or deliberative democracy may serve as a legitimating discourse that hides the administrative quality of participation rather than a democratic doctrine guiding the work of the participatory networks.

The case of Finnish LEADER also highlights certain shortcomings of the deliberative democracy theory. It shows that the urge to reach consensus may rather encourage uniform ways of thinking than pluralism. It also suggests that even small groups that hold consensus dear may need coercive means to support it – means to stop those that do not share the values from getting in. Striving to reach consensus may also turn out to be destructive. It may lead to such forms of communication (e.g. group think) which force people to accept decisions that would be rejected in other circumstances (Cooke 2001, 108–20). As one of my informants (Interview 14) wisely put it, one starts easily to evaluate the personal character of those who disagree with us. A more constructive and pluralist way would be listening to what they have to say, accepting the occurrence of different views and even seeing them as vital for on-going debate. Rather than striving for a shared consciousness, the perspective of politics would allow us to accept different and conflicting ideas and interests. If we want to see LEADER as a new form of democracy that gives rural citizens a channel for political action we should not be afraid of political conflict. Instead, the idea of publicly addressed dispute and political representation (see Urbinati 2006; Pitkin 1967) might enable larger segments of rural population a means of taking part in decision-making process. Even in the field of rural development we might be able to find some relevant political questions that could make it possible to think of LEADER as a kind of pluralist polity.

References

Cooke, B., 2001. The Social Psychological Limits of Participation. In: Cooke, B. and Kothari, U., eds. *Participation the New Tyranny?* London: Zed Books, pp. 102–21.

European Communities, 1988. *The Future of Rural Society*. Commission communication transmitted to Council and to the European Parliament on 29 July 1988. Bulletin of the European Communities, Brussels.

Hobson, K., 2009. On a Governmentality Analytics of the 'Deliberative Turn': Material Conditions, Rationalities and the Deliberating Subject. *Space and Polity*, 13(3), pp. 175–91.

Hyyryläinen, T., 2007. Toimintaryhmätyö paikallisen kehittämisen metodina. *Maaseudun uusi aika*, 15(3), pp. 20–36.

Karhio, K., 2000. Paikallisen kumppanuuden pitkä tie. In: Hyyryläinen, T. and Rannikko, P., eds. *Eurooppalaistuva maaseutupolitiikka*. Tampere: Vastapaino, pp. 78–107.

Kull, M., 2008. *EU Multi-level Governance in the Making. The Community Initiative LEADER+ in Finland and Germany*. Helsinki: University of Helsinki, Department of Political Science.

March, J. G. and Olsen, J. P., 1989. *Rediscovering Institutions*. New York: The Free Press.

———, 1995. *Democratic Governance*. New York: The Free Press.

Newman, J. and Clarke, J., 2009. *Publics, Politics and Power. Remaking the Public in Public Services*. London: Sage.

Nousiainen, M., 2011. *Hallinta, osallistuminen ja toimijuus – tulkintoja suomalaisen LEADER-toiminnan poliittisuuksista*. Mikkeli, Seinäjoki: Helsingin yliopisto, Ruralia-instituutti.

OEDC (Organisation for Economic Co-operation and Development), 2008. *OECD Rural Policy Reviews: Finland*. Paris: OECD Publishing.

Palonen, K., 1993. *Politisointi, politikointi, politiikka*. Helskinki: Jyväskylän yliopisto, valtio-opin laitos.

———, 2003a. Four Times of Politics: Policy, Polity, Politicking and Politicization. *Alternatives*, 28(2), pp. 171–86.

———, 2003b. Politiikka. In: Hyvärinen, M., Kurunmäki, J., Palonen, K., Pulkkinen, T. and Stenius, H., eds. *Käsitteet liikkeessä*. Tampere: Vastapaino, pp. 467–518.

Phillips, N. and Hardy, C., 2002. *Discourse Analysis: Investigating Processes of Social Construction*. Thousand Oaks: Sage.

Pitkin, H.F., 1967. *The Concept of Representation*. Berkeley: University of California Press.

Pylkkänen, P., 2004. Alhaalta ylös ja osallistuminen. In: Pylkkänen, P., ed. *Suomen LEADER+ -ohjelman väliarviointi 2003*. Helsinki: Maa- ja metsätalousministeriö, pp. 74–95.

———, 2008. Maaseutupolitiikan arviointikäytännöt hallintateknologioina. *Sosiologia*, 45(3), pp. 213–32.

Pylkkänen, P. and Hyyryläinen, T., 2004. Mainstreaming of the LEADER method into rural development policies in Finland. *Maaseudun uusi aika*, 12(4), pp. 22–32.

Ramos, E. and Mar Delgado, M., 2003. European Rural Development Programmes as a Means of Strengthening Democracy in Rural Areas. In: Bell, Michael M., Hendricks, Fred T. and Bacal, Azril, eds. *Walking Towards Justice. Democratization in Rural Life*. Research in Rural Sociology and Development, Volume 9. Oxford: Elsevier, pp. 135–57.

Setälä, M., 2003. *Demokratian arvo*. Helsinki: Gaudeamus.

Urbinati, N., 2006. *Representative Democracy*. Chicago: The University of Chicago Press.

Uusitalo, E., 2009. *Maaseutu – väliinputoajasta vastuunkantajaksi. Maaseutupolitiikan itsenäistyminen alue- ja maaseutupolitiikan puristuksessa*. Mikkeli, Seinäjoki: Helsingin yliopisto, Ruralia-instituutti.

Vehmasto, E., Vuorio, H. and Lahtinen, E., 2003. *Toimintaryhmäjoukkueen pelikirja – itsearvioinnilla yhteispelitaidot kuntoon*. Helsinki: Maa- ja metsätalousministeriö.

Wade, P. and Rinne, P., 2008. *A Leader Dissemination Guide Book based on Programme Experience in Finland, Ireland and Czech Republic*. Helsinki: Rural policy committee.

Interviews

Interview 1 (2009) male LAG board member.
Interview 2 (2008) female LAG manager.
Interview 5 (2008) male LAG board member.
Interview 6 (2008) male LAG manager.
Interview 8 (2008) male LAG board member.
Interview 9 (2008) female LAG board member.
Interview 10 (2008) male LAG board member.
Interview 11 (2008) male LAG board member.
Interview 12 (2008) male LAG board member.
Interview 14 (2008) male LAG board member.
Interview 15 (2008) female LAG board member.
Interview 16 (2008) female LAG board member.

Chapter 7

LEADER and Possibilities of Local Development in the Russian Countryside

Leo Granberg, Jouko Nikula and Inna Kopoteva[1]

Introduction

Participatory action research (PAR) made a breakthrough in the international research field about 40 years ago. In developing countries it offered the promise of contributions to local development by researchers who would actively get involved in local activities. When studying the basic principles and objectives of the LEADER method, we can find similarities to participatory action research, even if the role of research is seen as smaller or even non-existent. In reality, academic professionals, with researcher education, often have a strong influence among rural developers in LEADER projects. This is one good reason to study the outcomes of such projects in the framework of action research. This chapter will present the evaluation research, which was carried out in connection to a development project 'Ladoga Initiative' (LI) in North-Western Russia. LEADER had never before been operated in Russia. In this pilot project (2011–2013) local actors applied the LEADER method to find out and to support small-scale initiatives in the vicinity of Lake Ladoga. The main question in the evaluation is how well this approach can be applied to the circumstances of rural Russia. Before moving to the process and the results of establishing a LEADER type of

1 The subject of this chapter is a project of European Union's ENPI programme (European Neighbourhood and Partnership Instrument), which was decided upon by EU's, north-western Russia's and eastern Finland's authorities. The two-year project was coordinated by the Ruralia Institute, Mikkeli, University of Helsinki and its evaluation research was done in collaboration with the Aleksanteri Institute. The Finnish Academy of Sciences supported field research work. This analysis is a contribution to the research programme 'Choices of Russian Modernisation' of the Finnish Centre of Excellence of the Aleksanteri Institute. The writers want to thank these European and Finnish Institutes for their support and especially Russian partners in the ENPI project. We are grateful to Laura Kauppila for checking the language. The chapter is an outcome of equal contributions by Leo Granberg and Jouko Nikula. Inna Kopoteva has contributed by writing about the background and the Russian projects and had an essential role as the organiser of our field work.

development effort in Russia, let us take a look at what LEADER is and how well it fits with the characteristics of participatory action research.[2]

LEADER

LEADER has been developed as a central method of rural development of the European Union's Common Agricultural Policy. The goal is to channel funding for small projects on the basis of local initiatives, which are supposed to facilitate development in the region in question. The funding instructions and the procedure of steering funding vary country by country. What is common in all countries is that funding is based on a local development programme, accepted by the local action group (LAG). The LAG is constructed on a partnership principle, representing, ideally, the local rural population. It can be formed, for example, from persons in administration, entrepreneurs, NGOs and individual local people, with a defined quota for each group. LAGs are resourced by national authorities and the EU. Local actors are called to form projects and to apply for funding from LAGs.

In this form, LEADER has many benefits, among others it supports small-scale projects which often miss funding in spite of being of high quality. Because projects are locally adopted, local people and project leaders are strongly committed to implementing them. Projects are also supposed to increase social capital, and therefore facilitate future development in various ways. The variation between countries is rather great when it comes to implementing LEADER principles. Project funding has often been criticised for being short-term and leaving continuation in the hands of Fortuna (a new sponsor). LEADER has also been criticised for establishing a new social group of 'developers' and leaving aside marginalised groups in rural areas (see Csurgó and Kovach in Chapter 4). This is partly true, because there is a clear stratification of actors in the LEADER. At the bottom are local activists, who run the projects, and at the mid-level there are national experts and decision-makers, who filter information and channel the monetary flows from the EU and state budget to the lower levels. At the top are international experts and decision-makers who possess and process critical information about the LEADER. However, the approach can also be characterised as a social movement which is reflexive to its surroundings and has reached the capacity to drive its own interests in the political field (OECD 2006).

LEADER and Goals of Participatory Action Research

Reason and Bradbury (2002, 1) specify five characteristics for action research:

2 'Action research is best seen as an emergent, evolutionary and educational process of engaging with self, persons and communities which needs to be sustained for a significant period of time' (Reason and Bradbury 2002, 12).

1. Participation and democracy
2. Working for practical outcomes
3. Creating new forms of understanding
4. Human flourishing as the basic aim of action and
5. Creating an emancipating process; producing not only knowledge but new abilities to create knowledge.

It seems that these characteristics have been adopted in the LEADER 'ideology'. LEADER includes the participation of various stakeholders in practical efforts as well as facilitating the participation of local actors in local development. The question of democracy is more complicated, because LEADER adds a new level of action and underlines partnership relations between the main groups of actors which also challenges traditional democratic structures and may lead to the exclusion of some local groups and individuals from crucial decisions.

LEADER projects are working for practical outcomes. Also, LEADER assists in sustaining local know-how and tacit knowledge. It depends on the circumstances, the extent to which different forms of knowledge interact and create new forms of understanding. Human flourishing is also one of the basic objectives in LEADER. This is clarified by comparison with measures of agricultural policy in the Common Agricultural Policy (CAP) in which (agricultural) production is the primary aim and other aims, e.g. connected to environmental values, producers' incomes, or welfare, are only of secondary importance. And finally, LEADER may contribute to the creation of an emancipatory process, producing not only knowledge but new abilities to create and use knowledge. This takes place when short-term projects lead to somewhat more permanent elements of development, e.g. when establishing public places for local communication, constructing institutional and educational structures, etc.

From Centralised Exogenous Towards Localised Neo-Endogenous Development Model

Changes in the models of rural development policy have emphasised the role of LEADER-type activity in rural development efforts. Beforehand, rural development was based on exogenous development, where new industries and their associated technologies, skills and patterns of working were imported into rural areas in order to overcome problems of marginality and backwardness. The aim of exogenous development was to integrate rural areas into the national and international economy (Murdoch 2000, 412). This was also true in the Soviet Union. There the modernisation of rural areas was based on the ideals of Soviet modernisation, where urbanisation and scientific-technological progress were the key elements. In practice this meant conscious efforts to create urbanised villages based on industrialised agriculture.

The Western European version of the exogenous development model had a number of weaknesses e.g. an over-reliance on state support, a dependence on large-scale firms often operating in single sectors, and a consequent marginalisation of small-scale, local entrepreneurs (Day and Hedger 1990; Lowe et al. 1998). In the West state-centred exogenous development was already being challenged in the 1970s by the idea of endogenous development, which emphasised locality: local actors, local resources and local initiatives (Murdoch 2000; Terluin 2003). Local actors were encouraged to take responsibility for the design and execution of development strategies. The down-sides of an endogenous approach are the domination of powerful actors (with the marginalisation of others) and the danger of local passivity. Besides, endogenous approaches neglected the impact of outside influences on local development, whether from globalisation, foreign trade or other actors (Murdoch 2000; Lowe et al. 1998; Ward et al. 2005).

Both exogenous and endogenous models of development represent ideal typical models. In practice development strategies include elements of both models, described with the term 'neo-endogenous approach'. This focuses on interaction between the local arena and the wider political, institutional, market and natural environment (Ward et al. 2005, 5; Ray 2006, 286; see Chapter 1 in this volume). In Russia, rural development policies have not departed from the exogenous model, even if the importance of entrepreneurship and individual activity are recognised by some political statements. Rural development is still both centralised and hierarchical, and actors at lower levels are strongly dependent on the interests of the upper levels and the decisions made there.

Action Research – Research in the Service of Development

Because the LEADER approach applies some of the basic ideas of action research, it is worth taking a look at the aims and uses of action research. The tendency of academic research has been, from time to time, to pay attention to the relation between the methods and practical goals of social research. A fresh debate was initiated by sociologist Michael Burawoy, who reintroduced the concept of public sociology.[3] According to him, it is 'a sociology that seeks to bring sociology to publics beyond the academy, promoting dialogue about issues that affect the fate of society …' (Burawoy 2005, 104). In many senses, Burawoy's challenge was already met 30 years earlier by Third World researchers, who developed action research. Its objective was very much the same as that of public sociology. According to Reason and Bradbury, 'action research is a participatory, democratic process concerned with developing practical knowing in the pursuit of worth-while human purposes, grounded in a *participatory* worldview which we believe is emerging at this historical moment' (2002, 1). Participatory Action Research starts from somewhat different premises than Borawoy's public sociology. For him the public sociologist acts as a spokesperson to people; the

3 First introduced by Herbert Gans (1988). See also Agger (2000).

local and often counter-public, whereas PAR aims at empowering people, at making them researchers and analysts of their own lives.

The theoretical roots of participatory action research partly go back to the work of social psychologist Kurt Lewin, who already in 1946 used the concept of action research and to John Dewey, whose followers have developed educational action research. Another line of the practical roots goes to black communities in the United States during President Johnson's administration in the 1960s, where the participatory method was used for the 'War on Poverty' (Fals Borda 2002, 29; Nousiainen 2011, 63). A similar approach to local development in European circumstances was later adopted for Finnish village research in the 1970s (Rouhinen 1981; Hyyryläinen 1988; on critical evaluation, see Oksa 2012). Village research was an early component of rural development work before the establishment of rural development policy.

In a more general sense, Fals Borda and his colleagues felt there was a need for convergence between popular thought and academic science. For reorientation, science should be able to solve three broad challenges (Fals Borda 2002, 28–31), which concern 1) status and legitimacy of 'popular knowledge' in relation to scientific knowledge; 2) mechanisms or links between practice and theory: where should one start and what are the mediating steps or links between the two, and 3) the question about the status of those people/communities that we study; are they only 'objects' of study or can they also be granted some subjectivity?

In the case of Russia, the first challenge is obviously the ability to differentiate very general and often abstract explanations of disruption in local development from local and very concrete observations, stemming from every-day experiences (such as 'passivity' or 'kolkhoz mentality') and generate a new interpretation based on critical evaluation of both sources of knowledge. The second challenge concerns the incongruity between general, global models and local conditions: what are the historical, economic or sociocultural features that limit or modify the applicability of a certain model or pattern of development? The third challenge concerns the relationship between the researcher and the people or community that is studied. In action research, the researcher and the community are partners, aiming at social change, and not just passive informants. At the end of the day, the study should produce knowledge about mechanisms of change (how questions) and barriers to change (why questions).

Experimenting the LEADER Approach in Russia

Rural Russia

After the Soviet period, people in rural Russia have struggled over 20 difficult years for a minimum livelihood and to save their existing organisations and safety networks. The three main phases in transition were the time of privatisation (1989–1994), the reorganisation of the food system (1994–1998), and the recovering state agricultural and rural policy (1999–). A fourth new period has started from 2008 with national priority programmes, including sub-programmes for agricultural and rural development as well as increased resources for this sector.

In spite of favourable economic development in Russia, the rural standard of living is low when compared with the average in the Western European countryside or when compared with the standard among the Russian urban population. Budgetary security is weak in most rural municipalities, and the severe deterioration of infrastructure and institutions (social, health and cultural services) in villages is a long-term problem (Nefedova 2011). The passivity of rural populations is a claim often made; however, the experiences from our project do not give univocal evidence for such a notion. Regional differences in activity are significant and the causes for perceived passivity also differ according to social position, age and other criteria.

In agricultural policy, some steps towards family farming were taken in early 1990s. After that the main priority reverted to large-scale farming, where private companies took the place of state farms. During the first decade of the twenty-first century, investments in agriculture and in agricultural land have started to flow from national and foreign sources. Agri-business has rapidly grown in southern and central Russia (2011.) Private plots of the farming population have still maintained their decisive role in supplying many food products for Russian consumers. Forests have maintained their value for local entrepreneurs. Some of the new state activities in recent years have concerned debt arrangements for large farms and the creation of a credit system for agriculture, improvements in farming machinery (leasing arrangements), protection of domestic production, and the strengthening of food supply chains in different ways. Also improvements in rural living conditions in general are now mentioned in political programs which support entrepreneurship, improved housing conditions and health services among others.

Previous studies (Nikula et al. 2011; Granberg and Nikula 2006; Kulmala 2007) have testified that the key problems in rural areas in the Ladoga region are:

- Reduction of employment opportunities
- Outmigration of the younger generation from the countryside and an ageing rural population
- Minor or no support from local authority to local development efforts, because of very limited resources
- Few or no associations at the local level
- Small groups of active people
- Local governance is under constant change because of administrative reforms.

There are similarities and differences to the EU countryside in north-western parts of Russia. Basically similar demographic trends exist. Compared to new EU member states – some of which were former Soviet States – similarities are stronger still. In general, however, social problems in rural Russia are deeper and structural problems harder to solve than in the EU (Nikula et al. 2011). Russia is a large, economically diverse federal state with three different governmental levels (federal, regional and local level), each of which has its own interests. This fact means that the preconditions for a developmental model, which emphasises partnership and equal rights, shared goals and resources, differ necessarily from that envisioned in the Western European contexts. The efforts for neo-endogenous development at the local level face serious challenges in Russia, due to its traditionally highly centralised administrative culture. Other difficulties in implementing a neo-endogenous model in Russian rural areas are related to the above mentioned problems of weak civil society and the dependency of local actors on regional or federal leaders. Concretely, there are no equal partners at the local level, but two weak partners (local administration and local people/NGOs) and one better resourced partner (business) which does not necessarily wish to carry out and finance the development efforts.

Regional Location of the Ladoga Initiative

The development project 'Ladoga Initiative' was carried out in 2011–2013 on the lakeside of Ladoga. Three districts took part in the project, two from the Karelian Republic and one from the Leningrad region. The joint project between Finnish and Russian partners organised knowledge transfer based on Finnish experiences from LEADER practices. The Ladoga Initiative applied this method as far as possible, constructing short-term groups to support, choose and steer activity in small local projects, so-called mini-projects, which were proposed and organised by local villagers.

The three pilot regions are different: Pitkaranta is an industrial district, where mainly mining, forestry and the pulp and paper industry are developed. The Olonets District is a typical agricultural area, which still had nine large agricultural enterprises at the beginning of 1990s but only five continue their activity currently. Since World War II, the two districts have a common history and a similar net of settlements. The Lodeinoe Pole District is situated next to the Olonets District, but belongs to the Leningrad Region (*oblast*). The industrial structure of Lodeinoe Pole is a combination of industry (mainly forestry), agriculture and also tourism. The three big monasteries are a key tourist destination and more than 100,000 tourists visit them every year.

Before the World War II, Pitkaranta was a part of Finland. At the end of the war people were evacuated from there to Finland and later the region was populated by immigrants from different regions of Soviet Union. Olonets has a relatively large Karelian population. The proximity of the border is a reason for tourism, some business contacts and opportunities for seasonal working in Finland, more so in Pitkaranta and Olonets than in Lodeinoe Pole. In broad terms, however, there are no signs of them becoming westernised or being clearly different from other Russian regions.

Rural areas in Lodeinoe Pole are less developed in economic terms than in Karelia. The network of settlements also differs from its Karelian neighbours: the

Karelian rural municipalities usually have up to 10 villages, but in Lodeinoe Pole District one rural municipality can have up to 65 villages, most of them being small villages (1–20 persons). Also roads between villages are in a worse condition in Lodeinoe Pole than in Karelia. An explanation might be that the Karelian districts are located in the southern part of the Karelian Republic, which is a historically well-developed territory in Karelia, and even nowadays the main agricultural producer in the Karelian Republic. These two districts are also the most populated and are located rather close to the Republican capital Petrozavodsk.

The Lodeinoe Pole District is located on the periphery of the Leningrad oblast. Saint Petersburg and her neighbouring municipal districts attract the most investments, while Lodeinoe Pole is left in the margin of economic development of the oblast. So, even if the three pilot territories are located quite close to each other, their dynamics of economic development clearly differ from one another.

As already noted, there are many differences of circumstances in Russia compared with EU countries. Socioeconomic differences are remarkable and the political context is very different as well. Russian regional or rural policy programmes do not include policy measures such as LEADER. In that sense, the Ladoga Initiative is challenging existing Russian local development politics. However, there have been several earlier foreign projects, some of which have had features similar to the Ladoga Initiative.

Practical Organisation of the Project

The Ladoga Initiative was the first project in Russia, which deliberately adapted the LEADER method. Because it was a pilot experiment without planned continuation, creating longer-term Local Action Groups (LAG) was not seen as being possible. Instead, in each pilot territory all the villagers would be members of Local Initiative Groups (LIG). LIG boards were based on the partnership of three stakeholders: one third came from local authorities, one third from enterprises and one third from local people or representatives of voluntary organisations. Three local coordinators took the responsibility for project implementation in their own pilot territory. Coordinators were local residents living in district centres, and their role can be characterised as managers of the LIGs. The introduction to LEADER philosophy was carried out by training the coordinators, and other actors, in four different training sessions. During the training sessions in Finland, coordinators learnt about the practical implementation of LEADER. Also, they got acquainted with the necessary paperwork for concrete projects as well as how projects can be implemented and how to work with documents. The main goal of the training was to learn:

- The LEADER philosophy
- Practice how to implement project on own territory
- How to divide tasks between coordinators and LIGs.

The next stage was an essay competition for rural pupils with the topic 'The future of my village'. The project coordinators announced the competition in the hope that creative essay writing would bring children and adults together to discuss the problems of their home village. The next step was conducting village meetings, which had several goals: a) announce the idea of the project, b) discuss villages' problems and needs, c) make a list of development priorities for each rural municipality and d) suggest some candidates for the LIGs. These meetings were an attempt to engage, activate, and empower the local community. Altogether 27 village meetings were arranged in the project area. The reaction to the meetings varied greatly in different villages, as well as the participation rate and the nature of proposals concerning the needs and goals of local development. However, participation was very active when discussing villages' problems and needs in order to compose a list of development priorities for each rural municipality. The last task in each village meeting was to nominate three candidates to the LIG board. There was again variation among villages in how this task was fulfilled: most chosen candidates were teachers and employees of local administrations, and in two village meetings no candidates were nominated at all. When the LIG was established, it was time to organise a call for local mini-projects to be funded and organised by local activists.

Reception of the Ladoga Initiative

We received clarifying comments from the local interviewees concerning the Ladoga Initiative project and its ideas. First they noted that village meetings are arranged quite often in the regions, but they differ from the meetings arranged in the Ladoga Initiative. Russian village meetings are arranged around an actual and important local issue, but not to discuss more generally the tasks and challenges of village development. The LIGs succeeded not only in discussing but also in finding compromises between different proposed project efforts.

Another remark was that there is no lack of initiatives at the local level, but a lack of interaction between inhabitants and the government. The government was blamed for shifting the burden of helping the local community to the shoulders of local entrepreneurs and leaving the central state responsibilities unfulfilled. Respondents also mentioned an imbalance between business, administration and people. Administration and business interact a great deal, but interact less with local people.

Even though there were some men in the meetings and some male project leaders, women were more active than men in all meetings. The explanation given for this was that men are busy either in the political arena or in making money. Small projects do not interest male entrepreneurs as such. An addition was made by a respondent that 'Luckily in the countryside there are honest business people, who are earning with their own work' (Interviews 2012). In any case, some entrepreneurs were involved in the project activity and others took part in complementary activities. In one village an activist explained her motivation to participate as wanting to see the village as prosperous as it used to be 20 years ago. Not for herself but for her grandchildren.

At the beginning many people did not believe in the Ladoga Initiative and local coordinators visited the villages several times to help with different issues. It was especially difficult for local activists to present a budget for a project proposal. Training in making applications was helpful and even more would have been needed. Projects established a local network to plan the project together and to support it. The number of participants grew during the process. Local initiative groups held meetings in Russia and in Finland. Two Finnish experts taught the group members during the project and the atmosphere was very good. The belief among interviewees was that local authorities would support planned activities and that companies would also support them in some ways (Interviews 2012).

Monitoring the Project

Evaluation Study

A characteristic problem in this kind of development project is the difficulty of including research in any form into the project frame. We tried to solve this with a separate sub-project, which was based on ideas of action research. The research team was led by Jouko Nikula and consisted of Inna Kopoteva and Marjo Lehtimäki.[4] Inna Kopoteva, as the coordinator, was involved in the process of forming the local initiatives. She collected data about the localities, followed the procedure of formation of local initiative groups and their activities, received information on the formation of bottom-up initiatives and on the practical experiences of social partnerships. An external expert conducted thematic interviews during two intensive field work periods in order to analyse the impact of this pilot project on local rural development.

The project setting also opened the possibility of testing the impact of differences in regional circumstances and in the systems of local governance in the two regions involved, Leningrad Oblast and the Republic of Karelia. At the time of writing (summer 2013), the remaining larger challenge for the research was to discuss the applicability of the LEADER approach to Russian circumstances. It is clear that funded sub-projects give only short-term and limited results and our plan is to continue evaluation to reach at least mid-term results. To summarise, five purposes of the evaluation study were 1) to assist the LIGs in the process of initiating societal change; 2) to monitor the procedures of constructing the LIGs; 3) to evaluate a project's short-term impacts and to discuss its potential longer-term

4 Jouko Nikula was the leader of the local initiative research team, Leo Granberg participated in the role of the director of the Ladoga Initiative project and Inna Kopoteva as the project manager, and Marjo Lehtimäki worked both as the external adviser and monitoring phase expert.

impacts on community development; 4) to analyse the obstacles and achievements of bottom-up initiatives and social partnerships in the regions and 5) to evaluate the applicability of the LEADER approach to local Russian circumstances.

Main Results from the Ladoga Initiative

Research was composed of participatory observation in 2011–2012, intensive field research including thematic interviews in March and October 2012 and focus group work with members of three LIG groups in March 2012. All mini-project leaders or alternatively some other representatives from the projects were interviewed in October 2012. Returning to the five characteristics for action research by Reason and Bradbury (2002, 1), we reflect the results concerning the situation in the Russian countryside against our earlier experiences from 2000–2008 (e.g. Nikula 2011) in the following five points.

First, participation is part of the project. Project activities reached new groups in local society, especially middle class women, such as teachers, but also many other local residents, for example entrepreneurs, culturally active persons, children and youth. The target groups included depended on the contents and organisation of the project. Large groups of people were certainly left outside of the activity. One reason for participation was the high cultural and social capital among teachers and culturally oriented persons. They have a wider social network than others and for teachers and librarians it is even a kind of a social norm to be socially active; to organise events and to educate children and youth, etc. An accumulation of project activity was remarked, because some activists had earlier participated in other projects with foreign funding.

Second, most projects had very practical aims, to build a play yard, renovate a club house and so on. at the same time, many of these projects were strongly oriented to establish some form of public space: building playgrounds or sports fields for children and youth, renovating a cultural club, or a local museum and rebuilding a communal *banya* (sauna). This is understandable in a situation where most of the former public places have been closed or demolished. Because of closures and privatisation actions as well as reorganisation of major employers, the local administrations have had very limited chances to maintain or rebuild such premises. These kinds of places are important for local people in giving them an opportunity to meet each other, to get information, discuss and debate common matters, and to devise solutions to their joint problems.

Third, some projects had pedagogical aims or concrete aims to strengthen regional identity by establishing a Ladoga brand and arranging a local market event. A project which strived to strengthen local identity was a photography project by local youth in one village in Olonets District. The idea was to collect photos and videos of the history of the oldest residents of the village and to prepare an exhibition and leaflets. In this manner the project initiators hoped to teach young people the history of their village and strengthen their local identity.

The project was too short to bring together different forms of knowledge or to create new forms of reflexive understanding. Therefore, the continuation of projects is a crucial question and it was one topic to be discussed.

Fourth, human flourishing is a difficult aim to be operationalised, because it is a long-term goal. LEADER-activity can help to create preconditions to promote local development (prosperous business, healthy environment, reduced poverty, etc.) and in that manner finally reach the goal of human flourishing. Bottom-up activity opens opportunities for such human flourishing which local people are able and willing to aim for. They get the chance to consider their own needs and wishes and to set on positive development processes. As in LEADER in general, all population groups cannot be reached and other, additional measures are needed.

Fifth, it is not the time to draw conclusions on emancipating processes. There are, however, signals of other activities in Russia, with partly similar features, such as using the idea of local initiatives. In Archangelsk region local projects are used for some years already in wider scale to find solutions for local problems in collaboration between local people[5] and facilitated by regional administration. It is wishful to move on, from short-term projects to a larger experiment of development.

An important mechanism to support positive steps towards the five above mentioned aims is networking. Project activity opens possibility for networking on different levels, inside and between the three municipalities around Ladoga as well as between Russian and Finnish actors, and this network produces social capital. The network covers rather distant places and has some potential connections to Petrozavodsk and Saint Petersburg among others. However, the fate of this network depends on the future of mini-projects and LIG organisations or some – as yet non-existing – similar initiatives which would offer the network a necessary framework to sustain its activity.

Obstacles and Achievements

One idea behind the project was the unique nature around Lake Ladoga, and the value of the lake itself for recreation and tourism. Ladoga is the largest lake in Europe, but access to the lake itself is mostly unorganised. It would seem that the population and industries in the shore regions would have a lot to gain in life quality and in economic terms, if they could benefit more from this uniqueness. Among local projects such a topic was not prioritised, however. Instead, the clearest unifying aspect in the project proposals was to establish some kind of public space where people could go and meet each other and first and foremost places for children and youth.

Russian bureaucracy caused some difficulties for the projects and EU bureaucracy caused some more. Currency transfers caused losses in the small

5 This information was received both from local level and from regional administration during the field research in September 2014 by Ann-Mari Sätre, Lia Kalinnikova and Leo Granberg.

budgets of Russian partners and accounting and auditing had an unreasonable role, causing painful delays in funding for local micro-projects. At the same time, the resilience of local communities and actors turned out to be of a high class. When money was lacking, some other solutions were invented: voluntary work and the collection of local funding covered a good part of the lack of funding – which also finally found its way to the local projects. For the future, however, in these kinds of small-scale projects one needs innovations on how to steer funding to local needs in a simplified manner and how to minimise the need for book-keeping and reporting without losing confidence in the actors and partners in such projects.

When comparing the Ladoga Initiative to premises of participatory action research, the Ladoga Initiative did not fully meet them. The main reasons for this are, at first, the fact that the major part of the local population played a rather limited role in the Ladoga Initiative, while quite few local activists had appropriate organisational, cultural and social capital to run the process. The second reason was the limited role that the research group played at local level. The main influence came from the project manager, whose role was coordinating and monitoring the administrative activity of local initiative groups. Perhaps the most important result of the project was the clear rise in the awareness of the local population about the possibilities to change things even with small financial resources. Therefore, we can say that the project had a clear activating role: through examples of successful projects, local people understood that they have an impact on the future of their village.

Russian society has specific features which complicate the implementation of LEADER-type projects (see Granberg 2007):

1. Russia has a long tradition of strong and centralised control and command by central authorities on local action. As Stephen Wegren, among others, has shown, the Russian agrarian population, however, has in certain historical times been surprisingly flexible and ready to adapt to new circumstances. (Wegren 2005, 213) The positive reception of the LEADER method proves similar flexibility among the non-agrarian rural population. Experimenting with the LEADER method in the Russian local context demonstrates that opportunities for local activity exist.
2. Patriarchal culture is a heritage from a patrimonial pattern of ruling system in Russia for centuries (Weber 1921–1922; Granberg and Nikula 2010). New agency has, however, some space in Russian local society in rural areas. For example the rising activity of women from the rural middle class in development efforts appears to be quite a similar tendency as experienced earlier in the Nordic and Baltic countries. In some cases, the generational change among local authorities seemed to contribute towards a cultural change on the local level.
3. Partnerships in Russia have mainly concerned interaction between administration and large-scale enterprises (Nikula et al. 2011). In the Ladoga Initiative, the experiment with models of partnership between local

people and both administration and enterprises have been successful. Some differences can be seen in the achievements between the three LIGs.

4. Networking is based more on personal relations and less on formal civic organisations than in Western Europe. This state of affairs continues and will probably cause problems with preserving and developing the achievements attained in the long run. There is a need for new interested organisations to mediate local interests, to defend the positive results achieved and to work on new initiatives as well as a continuity of local action and utilising funding opportunities.

As a whole, the fresh results of the Ladoga Initiative have been promising. Local actors have been much more active than the writers supposed on the basis of preliminary information and the common beliefs about negative values among rural population, such as passivity, a kolkhoz mentality and a consumer mentality. Even if Russian circumstances are somewhat difficult for local initiatives, at least the short-term experiences from the Ladoga Initiative seem to verify that it is possible to initiate and to support new local development. Partly, such efforts are supported by some positive steps in social and economic development in rural Russia, which act as counter-effects to the ongoing economic closures and outmigration of the rural population.

References

Agger, B., 2007. *Public Sociology: From Social Facts to Literary Acts* (Second edition). Lanham, MD: Rowman & Littlefield Publishers.

Burawoy, M., 2005. American Sociological Association Presidential address: For public sociology. *American Sociological Review*, 70(1), pp. 4–28.

Day, G. and Hedger, M., 1990. Mid Wales: missing the point. *Urban Studies*, 27, pp. 283–90.

Fals Borda, O., 2002. Participatory (Action) Research in Social Theory: Origins and Challenges. In: Reason, P. and Bradbury, H., eds. *Handbook of Action Research*. London: SAGE Publications.

Gans H., 1988. *Middle American Individualism*. Free Press. Glencoe, Illinois: MacMillan.

Geddes, M., 2006. Partnership and the Limits to Local Governance in England: Institutionalist Analysis and Neoliberalism. *International Journal of Urban and Regional Research*, 30(1), pp. 76–97.

Granberg, L., 2007. Rural paths in Russia. In: Heinonen, M. Nikula, J., Kopoteva, I. and Granberg, L., eds. *Reflecting Transformation in Post-socialist Rural Areas*. Newcastle upon Tyne: Cambridge Scholars Press, pp. 49–62.

Hyyryläinen, T., 1988. *Kylien kehittäminen luovuuden näkökulmasta*. Aluetieteen laitoksen tutkimuksia, Sarja B51. Tampere: Tampereen yliopisto.

Interviews, 2012. Interviews made by research team in March and October 2012. Several persons in Pitkaranta, Olonets and Lodonaje Pole.

Kulmala, M., 2007. Russian Society. In: *Russia of Challenges*. Helsinki: Ministry of Defence, pp. 9–31.

Lewin, K., 1946. Action Research and Minority Problems. *Journal of Social Issues*, 2(4), pp. 34–46.

Lowe, P., Ray, C., Ward, N. and Wood, D., 1998. Participation on Rural Development: a review of European experience. Newcastle upon Tyne: Centre for Rural Economy Research Report, Department of Agricultural Economics and Food Marketing, University of Newcastle upon Tyne.

Murdoch, J., 2000. Networks – a new paradigm of rural development? *Journal of Rural Studies*, 16, pp. 407–19.

Nefedova, T., 2011. Sel'skoye khozyaystvo v Rossii do i posle reform 1990-kh godov [Agriculture in Russia before and after the 1990s Reforms]. In: S. Artobolevskiy, ed. *Regional'noye razvitiye i regional'naya politika Rossii v perekhodnyy period* [Regional Development and Regional Policy of Russia in The Transition Period]. Moscow: MGTU Im. Baumana, pp. 88–107.

Nikula, J., Kopoteva, I., Niska, M., Butkeviciene, E. and Granberg, L., eds, 2011. *Social innovations and social partnerships in Finland, Russia and Lithuania Helsinki*. Aleksanteri papers 2011:1 [Online]. Available at: http://www.helsinki.fi/aleksanteri/english/publications/files/Nikula_Social_innovations.pdf. Accessed 20.8.2014.

Nikulin, A., 2011. Oligarkhoz as Successor of Post-Kolkhoz. *Vestnik RYDN, seria 'sociologicia'*, 2, pp. 56–68.

Nousiainen, M., 2011. *Hallinta, osallistuminen ja toimijuus – tulkintoja suomalaisen leader-toiminnan poliittisuuksista*. Mikkeli, Seinäjoki: Helsingin yliopisto, Ruralia-instituutti.

OECD, 2006. *The New Rural Paradigm: Policies and Governance*.

Oksa, J., 2012. Thirty Five Years of Uneasy Affair: Triangle of Research, Village, and Policies. In: Nikula, J. and Granberg, L., eds. *Traces of Peasantry and Post-Socialism – Researching realities with Ilkka Alanen*. Jyväskylä and Tampere: Sophi, pp. 114–31.

Oksa, J., 2011. *Oksa's self assessment regarding activities and objectives* [Online]. Available at: http://www.korvesta-ja-valtateilta.fi/korvestajavaltateilta/uutiskirje/2011/2–2011/9.php

Ray, C., 2006. Neo-Endogenous Development in the EU. In: Cloke P., Marsden T. and Mooney P.H., eds. *Handbook of Rural Studies*. London: Sage, pp. 278–91.

Reason, P. and Bradbury, H., eds, 2002 [2001]. *Handbook of Action Research*. London: SAGE Publications.

Rouhinen, S., 1981. A New Social Movement in Search of New Foundations for the Development of the Countryside: The Finnish Action-Oriented Village Study 76 and 1300 Village Committees. *Acta Sociologica*, 24(4), pp. 265–78.

Terluin, I.J., 2003. Differences in economic development in rural regions of advanced countries: an overview and critical analysis of theories. *Journal of Rural Studies*, 19, pp. 327–44.

Ward, N., Atterton, J., Kim, T-Y., Lowe, P., Phillipson, J. and Thompson, N., 2005. *Universities, the Knowledge Economy and 'Neo-Endogenous Rural Development'*. Discussion Paper, Series 1, November. Newcastle upon Tyne: Centre for Rural Economy, University of Newcastle upon Tyne.

Weber, M., 1978 [1921–1922]. *Economy and Society* (volume I and II). Roth G. and Wittich, C., eds. Berkeley: University of California Press.

Wegren, S., 2005. *The Moral Economy Reconsidered.* New York: Palgrave Macmillan.

Chapter 8
Questioning the Gender Distribution in Danish LEADER LAGs

Annette Aagaard Thuesen and Petra Derkzen

Introduction

The aim of this chapter is to analyse and discuss the gender balance in Danish local action groups (LAGs) against the background of a reform in the LAG set-up aimed to secure input legitimacy. LAGs have become well-known in the domain of local rural development ever since the start of the European Union's Community Initiative LEADER in 1991. Since the 1990s, the emergence of a broad notion of integrated and sustainable rural development led to 'values about the active participation of stakeholders' coming to play a more distinguished role' (Greer 2005, 120). 'All of these approaches circumvented the top-down governmental approach associated with the traditional agricultural policy community' (Murdoch 2006, 174). Traditionally, the agricultural policy community has been male dominated. More local and informal decision-making has been thought to stimulate the entrance of previously excluded groups such as women. This chapter focuses on the gender distribution in LAGs because LAGs have become firmly established decision-making bodies for local rural development.

The LAGs can be seen as local expressions of a shift from government to governance that has taken place in European rural development policy as well as in many other policy areas with the objective of enhancing efficient and inclusive policy delivery at a local level. Despite the existence of affirmative gender policies accompanying the establishment of LAGs or other rural partnerships, concern still exists around the level of inclusion of previously disadvantaged groups in politics and decision-making, such as women. One of the factors which has been identified as hampering the inclusion of more women is the structural and cultural conditions of these new forms of governance; which are, as Pini (2004, 1) argues 'more of the same' in terms of gender inequality. In Denmark, concerns around accountability and representation have guided the Danish Ministry of Food, Agriculture and Fisheries to change the entry and selection methods for the LAG establishment from an appointment structure into elections in order to improve input legitimacy (see Chapter 5). Input legitimacy refers to the democratic legitimacy of the starting conditions of the decision-making entity; who decides on who can be a member and how are members accountable to the public so the public can control those that govern. These changes were not aimed at improving the gender balance per se, but were aimed at creating better access to the LAGs in general. Nevertheless, they have been the kind of structural changes that different authors have been calling for (Pini 2004; Bock 2004; Derkzen and Bock 2007). Hence, in the light of the reform of the access procedures, our aim here is to critically examine the LAG composition. Unlike previous, mostly qualitative case studies, this analysis is based on quantitative data representative for all Danish LAGs. This data goes beyond general figures on numerical counts and reveals the background and positions of LAG members. We ask the following question: 'Did the reform of the LAG organisation (the change from an appointment structure to an electoral structure) have any effect on the gender balance and what might be the future impact of this reform?'

Governance and Democratic Legitimacy

The new rural governance practice parallels an increasing body of literature addressing these changes theoretically. Governance theorists have described a shift from government to governance in which, on a more decentralised local level, decision-making increasingly takes place through multi-stakeholder platforms. In contrast to 'government', 'governance' therefore indicates a *pluricentric* rather than a *unicentric* approach to governing which moves the analysis away from a state-centric approach (Rhodes 1996; Heffen et al. 2000). It is argued that governance implies an increased importance of networks as the principle means for social coordination (Sorensen and Torfing 2003), in which 'hierarchy or monocratic leadership is less important' (Van Kersbergen and Van Waarden 2004, 152). Moreover, the governance literature has put emphasis on *processes* that highlight the negotiation, accommodation, cooperation and formation of alliances that occur. All in all, many have evaluated the perceived shift towards decentralisation and broader participation positively. Governance networks, or in European rural development jargon, 'partnerships' are seen as being capable of helping governments to deal more effectively with increased complexity and interdependency (Klijn et al. 1995; Rhodes 1996; Goodwin 1998; Bang 2003; Murdoch 2006). The new structures are said to improve the inclusiveness of decision-making towards previously excluded groups:

> Area based programmes are frequently presented as a means of addressing civic exclusion, both through the inclusive nature of the partnership structure, and through the local nature of the partnership, which is perceived to allow greater access to excluded groups than centralised policy. (Shortall 2004, 113)

However, studies which have focussed on the inclusion of women show that women hardly profit from the implementation of new rural governance modes (Bock 2002; Derkzen and Bock 2007; Little 2002; Pini 2004, Pini 2006; Shortall 2002). This is the case, even despite the existence of affirmative gender policies which often accompany the set-up of these governance networks in national as well as EU frameworks (CEC 2005; MFAF 2009). Thus, questions about legitimacy appear.

First of all, following gender studies, the representativeness of these partnerships is often seen as problematic. Representatives are often invited by a higher level of administration that turns to well-known established interest organisations and elites and often favours retired middle class white men (Bock and Derkzen 2003; Pini 2006; Shucksmith 2000; Thuesen 2010; Woods and Goodwin 2003; Young 2000). Secondly, the multi-stakeholder nature of the partnerships, in which the private sector and civil society organisations are also present next to the public sector, holds the danger that the boundaries between these sectors become blurred (Stoker 1998). Thirdly, such partnerships can create problems of accountability. They are basically an arena of decision-making outside the elected institutions of representative democracy which renders these governance networks only moderately accountable as a whole (Derkzen et al. 2008). It therefore induces the questions: who can participate and who are represented? Is it only benefiting the usual suspects; the local elite who 'can increase their power while at the same time denying it to already marginalised and excluded groups in the community' (Lawrence 2004, 10)? Based on a case study in rural Australia, Pini (2006, 404) concludes that the promise of power redistribution induced by language of 'partnership', 'networks' and 'participation' is a fallacy. Not only are women still hardly present, when moving beyond 'body counting' she shows the gendered binaries of masculinity and femininity inscribed in the governance process. Gender in rural governance, therefore, needs our attention. The more so because in rural development it seems that partnerships/networks are becoming an established method of decision-making; here to stay. Overall, there is therefore a tendency towards more network governance, which has been going on for some time and has both positive and negative implications and opportunities for women.

Outline of the Chapter

The next section explains the methods used. The following section introduces LEADER policy, its principles and changes in European rural development and fisheries policy as well as an in-depth explanation of the institutional changes in the LAG set-up in Denmark. Then, we move to the empirical material and show data on gender distribution among LAGs in Denmark. We show quantitative material from a questionnaire to all board members and discuss the gender balance and the background of the board members. Subsequently, in the following section, we discuss input legitimacy and the potential future impact of the reform of access procedures from a theoretical point of view (Mansbridge 2003; Saward 2006) and relate this to the findings. The final section offers a conclusion.

Methods

Data collection was undertaken by means of an electronic questionnaire sent to all 704 LAG board members in Denmark following a full census approach. This happened during the period from June 13 to August 1, 2008. 454 board members responded, constituting a fairly good response rate of 65 per cent. The full census approach leaves us in a situation resembling selection by self-selection instead of random selection. Selection by self-selection implies that the respondents in the sample are different from the rest of the population in one way, namely that they found reason to participate in the survey. Since we cannot say to what degree our sample resembles the population – that is whether it is biased or representative – we tried to evaluate this by making a comparison between sample and population. Nothing indicates that the data was biased regarding variables such as gender, role on the board, LAG type, municipality type and region. Even so, it is not possible to ignore the fact that for example non-respondents might have had a less positive view on the role of the LAGs.

The questionnaire sought data on the board members' experience with LEADER from previous programming periods, socio-economic characteristics of board members and accountability circumstances in the form of who the board members represented on the board. It also included questions concerning board members' motivations for joining the board and the amount and type of work they had put into the board, their view on the main strategies to follow as well as on the processes and results so far. In addition, questions about cooperation with the municipality, the region and the managing authority in the Danish Food Industry Agency and the Network Unit were included in the questionnaire. The data in this chapter stems from those parts of the questionnaire where a gender division is relevant.

Background: LEADER Policy – Its Principles and Changes

What is LEADER?

The central element of the LEADER method is the formation of LAG partnerships with representatives from the public, private and voluntary sector. The LAG partnership is supposed to decide on a local development strategy and implement the strategy through project grants to local project holders. In all, the LEADER method consists of seven concepts: area-based local development strategies, partnerships in the form of LAGs, bottom-up strategy, implementation of innovative strategies and cooperative projects, cross-sectoral approach, and networking of local partnerships.

During LEADER I (1991–1993), LEADER II (1994–1999) and LEADER+ (2000–2006) LEADER was a community initiative. This situation changed in

2007, when LEADER was mainstreamed into the rural development programme (RDP) and fisheries programme (FP) 2007–2013 (CEC 2005; CEC 2006a).

Institutional Changes in the Organising of Danish LAGs

The number of LAGs rose from 12 to 57 as part of the start of the RDP and FP 2007–2013 in Denmark and the number of persons active on the LAG boards rose from less than 200 to a little more than 700 persons (Thuesen 2010; Thuesen and Sørensen 2009). 39 LAGs work strictly within the RDP, 12 LAGs work both within the RDP and FP and six LAGs work strictly within the FP. The organisational set up was also changed in 2007 since before 2007, LAG board members were appointed not elected, and standards for entering Danish LAG boards were unclear. In addition, in most places there were no extra members of the LAG beside the board members. In contrast to this, in the programming period 2007–2013, it was compulsory for Danish LAGs to be organised as associations with open/free membership for all interested people over 15 years old living in the LAG area. LAG board members must be at least 18 years old and represent four groups: 1) local citizens, 2) local enterprises and trade organisations, 3) local associational life (nature, environment, culture, citizen, and leisure associations), and 4) public authorities. After 2007, board members are elected at annual general assembly meetings for two year periods (DFIA 2007a; DFIA 2007b).

The change from an appointment structure to an associational structure aimed to secure better input legitimacy for LAG decision-making in improving access and influence procedures by first, using the principle of election instead of appointment and second, by broadening the passive participation of inhabitants through associational membership. In the following section we will show our survey data concerning the composition of the LAGs.

Empirical Material: Gender Distribution on Danish LAG Boards

Distribution of Gender

A small proportion of women

Danish LAG boards have between seven and 21 board members and the average number of board members is 12 (Table 8.1). One of the guidelines from the ministry for the composition of the LAG boards in Denmark has been to aim at an equal gender distribution on the LAG boards. In spite of the formal guidelines, only 29 per cent of the LAG board members are women. In actual numbers, according to the population/email list, the lowest number of women on the LAG boards in June 2008 was zero and the highest number was eight.

Table 8.1 Gender Distribution of all LAG Board Members (N=454) and Gender Distribution of LAG Board Members Calculated by LAG Type (N=454)

	All LAG board members	LAGs of fisheries	LAGs of rural areas	LAGs of fisheries and rural areas
	%	%	%	%
Men	71	87	69	70
Women	29	13	31	30

The share of women shows a downward tendency, the more the LAGs have to do with the FP. Thus, LAGs acting strictly within the RDP have 31 per cent women on the board, LAGs acting in both the RDP and FP have 30 per cent women on the board and LAGs acting strictly within the FP have 13 per cent women on the boards. The FP has only included LAGs since the current programming period, and therefore, in some areas, it is the first time they have a LAG.

The share of women shows an upward tendency; the more the municipality is urbanised/centrally placed in Denmark (Table 8.2). This corresponds well with the fact that women make up a higher share of the population in these municipalities than in the outskirt municipalities of Denmark, but it could also indicate a more 'modern' lifestyle in the centrally placed municipalities, with women having a higher propensity to be involved. The differences are, however, not that large.

Table 8.2 Gender Distribution of LAG Board Members, Calculated by Municipality Type (N=454)

	Men	Women
	%	%
Outskirt municipalities	72	28
Rural municipalities	70	30
Intermediate municipalities	66	34
LAGs crossing municipality types	79	21

We can compare the current gender distribution with the pre-reform distribution in Denmark, for studies with similar percentages see (Bock and Derkzen 2003; Pini 2006, 407). We know from former research on the gender distribution of Danish LAGs in the start-up period of LEADER+ that female membership was low in this period, too (Thuesen 2003). The average share of women was 26 per cent. Eight out of the 12 LEADER+ LAGs had a share of women on the board below 25 per cent. Even though the current gender distribution is a bit less skewed, the 2007 reform did not at first sight have any significant effect on the gender distribution. The result is, however, bigger if we leave out the fisheries LAGs, which have the lowest share of women on the boards.

Other relevant political bodies can put the LAG board gender distribution in perspective. The Danish monitoring committees for the RDP and the FP have electronic member lists available on the web. These lists show that 52 per cent of the members of the monitoring committee for the RDP 2007–2013 in Denmark are women (DFIA 2009a). For the monitoring committee covering the FP 2007–2013 the share of women is only 14 per cent (DFIA 2009b). We can also compare the gender distribution of the LAG boards with other relevant elected bodies. We know that the share of female politicians on the municipal boards in Denmark is 27 per cent (Kvinfo 2009) and the share of female politicians in the European parliament is 35 per cent (Fifty-Fifty 2009). The LEADER+ Observatory Contact Point has collected data (monitoring indicators) from the former LEADER+ period via the national/regional managing authorities (LEADER+).[1] Here, the indicator of 'less than 50 per cent women' on the boards were much higher than the indicator of 'more than 50 per cent women' on the boards. Thus, the gender distribution of Danish LAGs does not seem to be different from the other European LAGs from the LEADER+ period. Danish LAG boards thus seem to be very much on the average level concerning gender distribution. Again, differences between gender distribution on rural and fisheries related boards are worth emphasising.

We now move on to look at the background of the LAG board members to go beyond mere body counting and the limited focus of numerical representation (Derkzen and Bock 2007; Pini 2006, 406; Shortall 2002, 168). Bock's (2004) research has shown that women's presence is 'not necessarily indicative of access to power, as those women who are members lack the access to knowledge, political experience, networks and institutional support that are available to male members' (Pini 2006, 397). However, it is not within the scope of this chapter to address the gendered subjectivities and meaning embedded in the LAG governance. Although, our ability to go beyond body counting is limited – given the quantitative nature of the data – the background data of the LAG board members shows interesting

1 There are differences in the reliability of the data between the countries, but the database seems to be the closest we get to a comparative summary of the gender distribution of European LAGs. Data is represented for the year 2004 in Annex 1. In seven countries, the largest shares of the LAGs had less than 25 per cent women on the board. In two countries, the largest shares of the LAGs had more than 50 per cent women on the board. Taken together, Finland, France, Luxembourg, Portugal, the Netherlands, Great Britain, Italy and Spain had LAGs with more than 50 per cent women on the boards. Greece was the country with the smallest share of women on the boards, since all 40 Greek LEADER+ LAGs had less than 25 per cent women on the boards (http://ec.europa.eu/agriculture/rur/leaderplus/midb_en.htm).

gender patterns which will be discussed in relation to the change in the access procedures from an appointment structure to elections in the discussion section.

General Background Characteristics

Age distribution

Women have a distribution above average in the age groups less than 50 years and men have distributions above average in the age groups over 50 years (Table 8.3). This indicates that the disproportionate representation may slowly change over time.

Table 8.3 Age Distribution, Calculated by Gender (N=452)[2]

	18–29	30–39	40–49	50–59	60–69	≥70	Total
	%	%	%	%	%	%	%
Men	2	8	18	40	30	3	100
Women	3	13	25	33	25	2	100
Total	2	10	20	38	28	3	100

Educational background

A higher proportion of female board members have a long cycle of higher education compared to men. This applies to 33 per cent of the female board members and 25 per cent of the male board members (Table 8.4). However, both male and female LAG board members in Denmark are very well educated since only about 6 per cent of the Danish population of roughly the same age group hold a long-cycle of higher education.

Main occupation

There is a slightly higher proportion of female wage earners (both public sector and private sector) and a slightly lower proportion of self-employed women compared to the data for men (Table 8.5). This can be explained partly by a relatively high representation of the agricultural and fisheries sectors as main actors in rural development and development of coastal areas. Accordingly, for example, a high number of farmers (mostly men) are registered under the category self-employed.

Even though the differences between women and men are moderate as to general background details such as age, educational level and occupation the data presented does to some degree contrast the argument by Bock about women's lack of knowledge. We now move on to look at the internal positions on the boards.

2 The shares do not actually sum up all the way to 100 due to rounding errors. This applies to Tables 8.3–8.7 and 8.9.

Table 8.4 Educational Background of LAG Board Members Calculated by Gender (N=454)

	Primary and lower secondary school	Upper secondary school	Vocational training	Short-cycle higher education	Medium-cycle higher education	Long-cycle higher education	Total
	%	%	%	%	%	%	%
Men	8	3	20	11	34	25	100
Women	5	5	9	15	32	33	100
Total	7	4	17	12	33	27	100

Table 8.5 Main Occupation of LAG Board Members Calculated by Gender (N=454)

	Wage earner (public sector)	Wage earner (private sector)	Self-employed	Senior citizen	Working at home	Under education	Other	Total
	%	%	%	%	%	%	%	%
Men	25	21	36	13	0	0	4	100
Women	29	25	27	11	1	2	5	100
Total	26	22	34	13	0	1	4	100

Positions on the Boards

Nominated positions

Women can be found in the highest proportion in the group 'local citizens' followed by 'local associations' and 'enterprises and trade organisations' (Table 8.6). Men, in contrast, can be found in the highest proportion in the group 'local associations' followed by 'enterprises and trade organisations' and 'local citizens'. Thus, all in all, fewer women are elected on nominated positions.

Table 8.6 Group for Which One Has Been Elected to the Board, Calculated by Gender (N=454)

	Public authorities	Enterprises and trade organisations	Local associations	Local citizens	Total
	%	%	%	%	%
Men	15	26	38	22	100
Women	11	20	29	40	100
Total	14	24	35	27	100

All LAG types seem to have one group that is clearly more weakly represented than the other three groups (Table 8.7). For LAGs of fisheries it is 'local citizens', while for the LAGs of rural areas and the integrated LAGs it is the group 'public authorities' that is most weakly represented. Since women are often elected for the group 'local citizens' this corresponds well with the low share of women on fisheries LAG boards (13 per cent) and the low share of board members elected for 'local citizens' in fisheries LAGs already mentioned.

Table 8.7 Group for Which One Has Been Elected to the Board, Calculated by LAG type (N=454)

	Public authorities	Enterprises and trade organisations	Local associations	Local citizens	Total
	%	%	%	%	%
LAGs of fisheries	26	32	35	6	100
LAGs of rural areas	14	22	36	29	100
Integrated LAGs	10	30	32	29	100

Positions held

The chairman of a LAG board in Denmark is elected at the general meeting, with the board subsequently deciding on a vice-chairman, treasurer and secretary. Given the smaller amount of women, they are on average slightly better positioned on the LAG boards than their numbers prescribe (Table 8.8). In all they hold 31 per cent of the primary positions. Women do however only hold the post as chairmen in about 26 per cent of the cases, whereas they hold the role as vice-chairman in about 33 per cent of the cases, the role as treasurer in about 31 per cent of the cases and the role as secretary in about 35 per cent of the cases. The figures indicate that even though women are to some degree externally excluded since the boards only consist of one third women representatives, female LAG board members in Denmark do not seem to be further internally excluded when entering the LAG boards (Young 2000) since they too obtain important positions. Even though they have a higher propensity to become vice-chairman, treasurer and secretary than to become chairman, they do obtain a share of the posts as chairman that almost equals their share of the total number of board members.

Table 8.8 Positions Held on the Board Calculated by Gender (N=454)

	Chairman	Vice-chairman	Treasurer	Secretary	Ordinary member of the board
	%	%	%	%	%
Men	74	67	69	65	71
Women	26	33	31	35	29
Total	100	100	100	100	100

Knowledge of other board members when entering

Women knew fewer board members before entering the board than men did (Table 8.9). This corresponds well with the fact that female representatives often represent the group 'local citizens' and thus are more loosely coupled to existing networks than the male representatives. Hence, the category of 'local citizen' seems to give women an entry point.

Table 8.9 Knowledge of Board Members when Joining as Board Members

	No one	1–2 persons	3–5 persons	6–10 persons	More than 10 persons	Total
	%	%	%	%	%	%
Men	11	29	37	18	5	100
Women	15	38	30	16	2	100
Total	12	32	35	18	4	100

To end the characterisation of male and female LAG board members we will emphasise that concerning women's and men's 1) level of commitment, 2) their assessment of the LAG goals and the results created during the first year, we cannot trace any clear differences. This is of course despite the fact that women only make up one third of the boards. Except from the differences already described about main occupation, group for which one has been elected, knowledge of other board members and the lower share of chairmen, we have only small indications and thus no striking data showing that women are structurally weaker positioned than men on the LAG boards (Table 8.10). This contrasts with previous research (Derkzen and Bock 2007; Pini 2006; Bock 2004 and Shortall 2002). Given the positions women take on the LAG boards, there is no evidence of women serving as the 'female other' within a strong gendered binary of masculinity and femininity (see for an example of the 'female other' Bock and Derkzen 2008, 276–7). However, we are well aware that the many ways in which exclusion can take place through gendered meanings and subjectivities are not visible from the quantitative data.

Table 8.10 Women's Relative Position to Men

Women relatively well positioned	Women not relatively well positioned
Slightly better educated	Represent 'local citizens'
Slightly younger	Knew fewer board members when entering
Hold positions on the boards that are a little higher than their share of board members	Lower share of female chairmen
	More seldom self-employed

Conclusion on this Section

As a first conclusion, when 'body counting', the data of this survey shows a continued gender imbalance in Danish LAGs. The average percentage of women (29 per cent) does not differ much from the average in the previous LEADER+ period. However, there is an interesting difference in the gender balance between the areas which implement the RDP and areas which implement the FP. The gender distribution becomes more skewed the more the LAG has to do with fisheries and especially the strictly fisheries LAGs show a lower proportion of women on their board. This corresponds well with the fact that the fisheries LAGs have the lowest share of the group 'local citizens' on the boards, since only 6 per cent belong to this group. As shown above, the largest part of the female LAG board members belonged to the group 'local citizens'. A second conclusion therefore indicates probably a small 'emancipatory effect' of the LEADER way of working of bottom up decision-making in relation to the rural development programme where the former LEADER+ period might have left its marks, something which is not the case in relation to the newly installed FP. The current system of associational membership and elections might well stimulate this further. We will elaborate on this in the next section.

Thirdly, when moving beyond body counting, we can see that behind the gender imbalance there are little differences as to the general characteristics (age, education, and occupation) of the board members be they men or women even though women are slightly younger, have a higher educational level and are more rarely self-employed. Tables 8.6, 8.7 and 8.8 show small but interesting differences pointing again to the 'door opening effect' the LAG might have in letting new groups enter the decision-making body. As mentioned, women are slightly more represented in the category 'local citizen' (Table 8.6). This category contains non-organised, independent citizens instead of nominated positions via organisational representation in other categories. Traditionally, organisations such as farmers unions, chambers of commerce and from these data maybe also fisheries organisations are often more male dominated. In the same vein, Table 8.9 seems to indicate again the independence of women's entrance from existing networks, persons or organisations in their report of low numbers of already known persons in LAGs and thus maybe the emancipator effect of the program although qualitative research following the decision-making process and *culture* over time is needed to elaborate on this tentative conclusion. We reflect more on these findings and the relation to input legitimacy in the next section

Discussion of the Gender Distribution in Relation to Input Legitimacy

In counteracting gender imbalance in decision-making networks such as LEADER LAGs there are two different policies at work. First, there is the general gender policy which tries to induce gender balance by affirmative action and rhetoric.

Second, in this case, there is a policy to improve input legitimacy which could also have an effect on the gender balance. Fair and equal access procedures (how to become a member of the LAG board) and influence procedures (how to influence decisions taken by the LAG boards) are important for input legitimacy. Although we have concluded that there still is an imbalance in the overall gender distribution, we will argue here, that the reform for the Danish LAGs to secure better input legitimacy might be more effective in counteracting gender imbalance than affirmative gender policies.

The General Gender Policy – Affirmative Action and Rhetoric

There are several possible reasons why fewer women than men end up on the LAG boards. Bock and Derkzen (2008) have outlined four different barriers to women's participation in rural policy making:

1. Women's position in rural society and their weak socio-economic and political integration.
2. A traditional gender ideology that underlines women's domestic responsibilities and civil and apolitical involvement in the community.
3. The dominance of agriculture and economy in the rural development discourse.
4. The lack of fundamental structural and cultural changes in new governance arrangements.

All these possible reasons point to the disadvantaged position of women in the public sphere. Mansbridge (2003, 99) states that disadvantaged groups might 'want to be represented by individuals who 'mirror' the typical experiences and outward manifestations of belonging to the disadvantaged group' (see Derkzen and Bock 2007, 200, for an example). In descriptive representation, the aim is to get more women in decision-making since women can represent best the interests of women, referring to both the visible characteristics of the group and to the shared experiences. Affirmative policies such as gender guidelines implicitly have some kind of assumption of descriptive representation underlying them. For example the Council regulation states:

> Member States and the Commission shall promote equality between men and women and shall ensure that any discrimination based on sex … is prevented during the various stages of programme implementation. This includes the stages of design, implementation, monitoring and evaluation. (CEC 2005, 10)

Also in Denmark, in both the departmental order and the standard rules concerning the foundation and running of LAGs, the gender distribution is addressed.

While our data confirms the skewed gender distribution for which these affirmative policies are made, our data also seems to break with the image of women as a disadvantaged group. From the general characteristics of the women *on* the LAG boards, we cannot conclude that they are distinctly disadvantaged in terms of resources and positions held.

Affirmative gender policies therefore, might not enhance the possibilities of women joining LAG boards. The groups who appear to be good candidates for affirmative policy to increase their representation, are those whose political participation was historically forbidden or hampered; which still works through today, only now in 'informal social political and economic structures rather than through law' (Mansbridge 2003, 110). Therefore, the existence of affirmative gender policies can express a lack of real interest to change the deeply inscribed structural and cultural ideas and practices which uphold that which the policy tries to counteract. Indeed it has been argued (Derkzen and Bock 2007) that rural women, representing their interests from a 'women's point of view' were seen as 'the other', the non-professionals in new rural governance networks. By thus emphasising the features by which inequality works, the possibilities for change become 'locked in' to dominance – resistance frameworks.

The Policy to Improve Input Legitimacy

The policy to improve input legitimacy on the other hand, has, we believe, more potential impact. This policy, which introduces new institutional rules and norms such as election instead of appointment, works directly into changing governance practices. It addresses, therefore, more directly the structural and cultural barriers which could possibly hamper the involvement of previously excluded groups like women (Bock and Derkzen 2008).

First, both the set-up of the association underlying the LAG board as well as the introduction of elections open up existing networks of cross-reference. The system of appointment usually works through existing networks of active people, formal or informal community leaders, governors or relevant organisations. Thus selecting board members by appointment naturally limits the potential 'pool' from which to choose. In contrast, selection through election and association has more potential to recruit outside the known networks, which may benefit previously excluded groups such as women, and categories which do not rely on nominal positions such as 'local citizens'.

Secondly, membership of the association potentially increases the (passive) participation and knowledge of the work the LAG does to a wider group of inhabitants (the LAG associations had up to 300 members at the time of the investigation). It offers the possibility of becoming acquainted with this type of rural development without being either a board member or a project initiator. It thus introduces a new type of involvement which might lead to other types of involvement in the future.

Thirdly, the new arrangements have more space for the *creative process* of representation (Saward 2006). This can be explained through Saward's concept of 'the representative claim'. In most theories of representation focus is on the representative and how he/she acts in the interests of the represented. The constituency is left somehow out of the picture as unproblematic assuming that their interests are more or less transparent. Saward argues that there is a dialectic relationship between the representative as 'claim maker' and the constituency as active in acknowledging, accepting or rejecting the claim. This view puts emphasis on the performative process of (being allowed) to represent.

> [W]ould-be political representatives, in this process of portrayal or representation of constituencies, *make claims* about themselves and their constituents and the links between the two; they argue or imply that they are the best representatives of the constituency *so understood.* (Saward 2006, 302, emphasis in original)

Saward's argument about the creative process of representation shows that for descriptive representation to work, the constituency – the women in the area – also have to identify with what the representative brings forward. Claim making from a women's point of view needs to meet with recognition and acknowledgement that this is a separate and a valid claim. In other words, it needs the women in the area to identify themselves as a separate and coherent constituency just because they are women. For representation in LAGs on the basis of gender only, the claim maker needs recognition from other women that they need political representation because they view themselves as being disadvantaged or being different.

Mansbridge argues that the benefits of enhanced deliberation of descriptive representation are highest in contexts of communicative distrust or un-crystallised interests. Since our data reveals that women are quite well positioned on the boards, there does not at first glance seem to be two opposing groups that cannot communicate. Neither do un-crystallised issues exist around the gender issue. In relation to the low share of women on municipal councils in Denmark Kjær emphasises that:

> There is no one in Denmark who is seriously discontented with the situation. Except from a few dedicated souls, no one is so indignantly dissatisfied that they raise the question. You don't here critical voices from the parties demanding a higher share of women – neither from the women in the party. (Kjær in Moustgaard 2009, 10, own translation)

So conditions for the stimulation of gender-based representation do not exist, nor does there seem to be conditions for successful claim making based on gender alone.

Rather, the theory of Saward points to the potential impact for improving the gender balance through the Danish reform to introduce elections and associational membership. Elections and the process of becoming elected involve a far more creative process than the appointment system does. Claim making is likely to be more explicit and articulated and to involve more people. Elections return periodically giving this process of claim making a chance to be refreshed and rehearsed. In contrast to the value laden aspects of the affirmative policies, the

Danish LAG reform opens the governance structure in a more gender neutral way. Through elections, new people can enter the rural development 'arena', not dependent on the appointment system, nor on nominal positions. The category of 'local citizens' therefore is a crucial category. Questions remain whether or not representatives in this category can influence decision-making in the same way as other representatives.

Conclusion

This chapter has uncovered the gender distribution of Danish LAG board members and contrasted the characteristics of female and male board members. It has also discussed whether the reform of the LAG set-up has had an effect on the gender distribution and assessed the future impact of the reform. Here we will now conclude on whether the views expressed in the literature section at the beginning of the chapter agrees with the Danish empirical data.

Securing the input legitimacy in Danish LAG's through the introduction of LAG associations, general assemblies and membership did not yet have a major effect on the gender balance in the Danish LAG's. The questionnaire results show a gender distribution, which does not deviate much from past patterns. The literature has stated that the new rural governance structures are just more of the same. This holds true as to the female representation on the board which makes up 29 per cent.

However, if we look behind the general 'body count', the situation seems to change. We have uncovered that Danish female LAG board members are slightly better educated, slightly younger and hold positions on the board that are a little higher than their share of board members except the slightly lower proportion of chairmen. At the other hand, women represent the group 'local citizens', know fewer board members when entering and are more seldom self-employed. The study does however show no striking data that women are structurally weaker positioned than men on the LAG boards and this contrasts with previous research (Bock and Derkzen 2008; Pini 2006; Bock 2004; Shortall 2002). Given the positions women take on the LAG boards, there is no evidence of women serving as the 'female other' within a strong gendered binary of masculinity and femininity. However, we are aware that ways in which exclusion *can* take place through gendered meanings and subjectivities are not visible from the quantitative data.

The Danish reform to secure input legitimacy for LAG decision-making is the kind of reform which various authors have called for (Derkzen and Bock 2007; Pini 2006; Shortall 2002). Moving from an appointment system to an election system opens the governance structure and – maybe over time – the culture of decision-making. The re-organisation in associations can open space for other constituencies to appear, and other representative claims to occur over time. Despite very moderate reflection of the impact of this reform in our data, the reform seems to be a good first step towards formalizing equal and open access opportunities for all at the place of former access procedures through more closed networks.

Annex 1

Table 8.11 Gender Distribution on LAG Boards in Europe, 2004

	Number of LEADER+ LAGs	Participation of women on the LAG boards		
		<25 %	25–50 %	>50 %
		%	%	%
Denmark	N=12	58	33	8
Austria	N=56	18	82	0
Belgium	N=20	55	45	0
Finland	N=25	4	84	12
France	N=140	54	16	30
Germany	N=148	47	49	5
Greece	N=40	100	0	0
Ireland	N=22	32	68	0
Luxembourg	N=4	0	0	100
Portugal	N=52	2	4	94
Sweden	N=12	0	100	0
The Netherlands	N=28	54	25	21
Great Britain	N=57	9	63	28
Italy	N=130	73	10	17
Spain	N=146	68	18	14

References

Bang, H.P., 2003. *Governance as social and political communication.* Manchester and New York: Manchester University Press.

Bock, B.B., 2002. *Tegelijkertijd en tussendoor. Gender, plattelandsontwikkeling en interactief beleid* [*At the same time and in between. Gender, rural development and interactive policy*].Wageningen: Wageningen University, p. 206.
———, 2004. *Gender and new rural governance: theory and practise of rural women's inclusion in policy making*. In: IRSA World Congress on Rural Sociology. Trondheim, July 2004.
Bock, B.B. and Derkzen P., 2003. *Vrouwen en reconstructiezaken. Deelname van vrouwen aan de besluitvorming over het reconstructieproces in Gelderland.* [*Women and reconstruction issues. Participation of women in decision-making on the reconstruction process in Gelderland*]. Wageningen and Utrecht: Wageningen universtiteit / Vrouwenalliantie.
———, 2008. Barriers to women's participation in rural policy making. In: Morell, I. A. and Bock, B.B. *Gender Regimes, Citizen Participation and Rural Restructuring*. Oxford: Elsevier, pp. 265–84.
CEC, 2005. *Council regulation No. 1698/2005 of 20 September 2005*. Official Journal of the European Union.
———, 2006a. *Council regulation No. 1198/2006 of 27 July 2006*. Official Journal of the European Union.
———, 2006b. *The LEADER approach – A basic guide*. Fact Sheet. European Commission.
Derkzen, P. and Bock, B.B., 2007. The Construction of Professional Identity: symbolic power in rural partnerships in the Netherlands'. *Sociologia Ruralis*, 47(3), pp. 189–204.
Derkzen, P., Franklin, A. and Bock, B.B., 2008. Examining power struggles as a signifier of successful partnership working: a case study of partnership dynamics. *Journal of Rural Studies*, 24(4), pp. 458–66.
DFIA, 2007a. *Vejledning om oprettelse af og tilskud til drift af lokale aktionsgrupper i landdistrikterne* [*Guidance on establishing and subsidizing the operation of local action groups in rural areas*]. Danish Ministry of Food, Agriculture and Fisheries/Danish Food Industry Agency.
———, 2007b. *Vejledning om oprettelse af og tilskud til drift af lokale aktionsgrupper i fiskeriområderne* [*Guidance on establishing and subsidizing the operation of local action groups in rural areas*]. Danish Ministry of Food, Agriculture and Fisheries/Danish Food Industry Agency.
———, 2009a. *Medlemmer af overvågningsudvalget for Landdistriktsprogrammet 2007-2013* [*Members of the monitoring committee for the Danish RDP 2007-2013*]. Available at: <http://naturerhverv.dk/tvaergaaende/eu-arbejdet/landdistriktsprogrammet-2007-2013/#c16053 > [Accessed 16.12.2014].
———,2009b. *Medlemmer af overvågningsudvalget for Fiskeriudviklingsprogrammet 2007-2013* [*Members of the monitoring committee for the Danish FP 2007-2013*]. Available at: <http://naturerhverv.dk/index.php?id=2893#c28776> [Accessed 16.12.2017].

Fifty-Fifty, 2009. *European Women's Lobby 50/50 Campaign for Democracy* [Online]. Available at: <http://www.5050democracy.eu> [Accessed 24 October 2014].

Goodwin, M., 1998. The governance of rural areas: some emerging research issues and agendas. *Journal of Rural Studies*, 14(1), pp. 5–12.

Greer, A., 2005, *Agricultural policy in Europe*. Manchester and New York: Manchester University Press.

Heffen van, O. Kickert W.J.M. and Thomassen, J.J.A., 2000. *Governance in modern Society: Effects, change and formation of government institutions*. Dordrecht: Kluwer Academic Publishers.

Kersbergen van, K. and Waarden van, F., 2004. 'Governance' as a Bridge between Disciplines: cross-disciplinary inspiration regarding shifts in governance and problems of governability, accountability and legitimacy. *European Journal for Political Research*, 43(2), pp. 143–71.

Klijn, E.-H., Koppenjan, J.F.M. and Termeer, K., 1995. Managing Networks in The Public Sector: a theoretical study of management strategies in policy networks'. *Public Administration*, 73(3), pp. 437–54.

Kvinfo, 2009. *27 procent kvinder i dansk lokalpolitik – hvad er problemet?* [*17% women in Danish local politics – what is the problem?*] [Online].<http://kvinfo.dk/2015/27-procent-kvinder-i-dansk-lokalpolitik-hvad-er-problemet> [Accessed 4 December 2014].

Lawrence, G., 2004. *Promoting Sustainable Development: The Question of Governance.* Plenary address XI World Congress of Rural Sociology. Trondheim, Norway, 25–30 July 2004 [Online]. http://www.irsa-world.org/prior/XI/program/Lawrence.pdf [Accessed 4 December 2014].

LEADER+, n.d. The Leader+ Monitoring Indicators Database (MIDB) [Online]. Available at: <http://ec.europa.eu/agriculture/rur/leaderplus/midb_en.htm> [Accessed 24 October 2014].

Little, J., 2002. *Gender and rural geography*. Harlow: Prentice Hall.

Mansbridge, J., 2000. What does a Representative Do? Descriptive Representation in Communicative Settings of Distrust, Uncrystallized Interests, and Historically Denigrated Status. In: Kymlicka, W. and Norman, W. *Citizenship in Diverse Societies*. Oxford and New York: Oxford University Press, pp. 99–123.

MFAF, 2009. *Order number 141 of 24/02/2009*, Danish Ministry of Food, Agriculture and Fisheries.

Moustgaard, U., 2009. En af tre er 'nok' [*One out of three is 'enough'*]. *NIKK Magasin*, 3, Oslo: Nordic Gender Institute.

Murdoch, J., 2006. Networking Rurality: emergent complexity in the countryside. In: Cloke, P., Mardsen, T. and Mooney, P.H. *Handbook of rural studies*. London, Thousand Oaks and New Delhi: Sage Publications, pp. 171–84.

OECD, 2006. *The new rural paradigm: Policies and Governance*. Paris: OECD Rural Policy Reviews.

Pini, B., 2004. *Masculinities and new forms of governance in rural Australia.* Paper presented at the World Congress for Rural Sociology in Trondheim/ Norway, July 2004.

———, 2006. A critique of 'new' rural local governance: The case of gender in a rural Australian setting, *Journal of Rural Studies*. 22, pp. 396–408.

Rhodes, R.A.W., 1996. The new governance: governing without government. *Political Studies, XLIV*, pp. 652–67.

Saward, M., 2006. The Representative Claim, *Contemporary Political Theory*, 5(3), pp. 297–318.

Scharpf, F.W., 1997. *Games real actors play – Actor-centered institutionalism in policy research.* Boulder, Colorado: Westview Press.

Shortall, S., 2002. Gendered agricultural and rural restructuring: a case study of Northern Irelan'. *Sociologia Ruralis*. 42(2), pp. 160–75.

———, 2004. Social or economic goals, civic inclusion or exclusion? An analysis of rural development theory and practice. *Sociologia Ruralis*, 44(1), pp. 109–23.

Shucksmith, M., 2000. Endogenous development, social capital and social inclusion; perspectives from LEADER in the UK. *Sociologia Ruralis* 40(2), pp. 208–18.

Sorensen, E. and Torfing, J., 2003. Network politics, political capital, and democracy. *International Journal of Public Administration*, 26(6), pp. 609–34.

Stoker, G., 1998. Governance as theory: five propositions'. *International Social Science Journal*, 155, pp. 17–28.

Thuesen, A.Aa., 2003. *'Det handler om at få en plan til at virke ...' Undersøgelse af de danske lokale aktionsgrupper i LEADER+*. Esbjerg: CFUL.

———, 2010. Is LEADER Elitist or Inclusive? Composition of Danish LAG Boards in the 2007–2013 Rural Development and Fisheries Programmes. *Sociologia Ruralis*, 50(1), pp. 31–45.

Thuesen, A.Aa. and Sørensen, J.F.L., 2009. *Danish Local Action Groups in Rural and Fisheries Areas 2008 – composition, activities and cooperation in the start-up phase.* Esbjerg: Danish Institute of Rural Research and Development.

Woods, M. and Goodwin, M., 2003. Applying the rural: governance and policy in rural areas. In: Cloke P., ed. *Country Visions*. Harlow: Pearson Education Ltd, pp. 245–62.

Young, I.M. (2000) *Inclusion and Democracy*. Oxford: Oxford University Press.

Chapter 9
LEADER LAGs: Neocorporatist Local Regimes or Examples of Economic Democracy?

Giorgio Osti

Introduction

The aim of this chapter is to understand in what terms LEADER is a case of local democracy. To this end, three questions are addressed. First question is whether the March and Olsen (1989) integration/aggregation dichotomy is useful for interpreting the issue or would a triadic model be more appropriate. Second one is, if the concept of 'local democracy' is really broad enough to include also the material conditions (income, time, accessibility) of participation in the most important areas of public life. Third question is how the Italian case can help in the endeavour. The two first questions concern the framework of the topic, and the last question is the object of analysis. The reference to March and Olsen's dichotomy is due to the international research project on LEADER that used it as its main theoretical basis (see other chapters in the volume). The question concerning broadness of participation relates to the debate on economic democracy. Italy fits well for researching these topics, because of its strong tradition of municipal action – that is a territorial level, which is very similar to the territorial concept of LEADER.

Integration, Aggregation and More: Toward a Triadic Model

With particular regard to the practical object of inquiry – LEADER as a case of bottom-up development – clarification should be made of the theoretical meanings of integrative and aggregative ways to determine the will of the people (Italian: *volontà popolare*) through institutions considered to be models of democracy. March and Olsen's (1989) idea was precisely to illustrate legitimated sets of rules useful for representing the petitions of people.

The aggregative pattern is an institutional type of governance that resembles a market: numerous independent actors negotiate their interests and achieve a substantial balance in the entire socio-political system. 'The traditions of aggregation define "the people" as a collection of individuals currently qualified to be treated as citizens' (March and Olsen 1989, 118). The central factor is exchange. The institutions are mainly concerned with the efficient allocation of resources amongst actors. The political leadership acts as a sort of mediator among contrasting interests. The role of the public sector is therefore very minor: it is required for control and for the distribution of very selective incentives.

The integrative pattern is another form of governance that recalls a community. The emphasis is on goods, values and destinies, which are deemed common and more important than individual interests. The crucial factor is a common cultural identity. 'The traditions of integration define "the people" as a group extending backward through history and forward through the future' (March and Olsen 1989, 118). A higher synthesis is possible, and the public institutions represent such an idea. The political leadership performs a moral role in that it interprets the community's history and future. Much emphasis is placed on the education of both politicians and citizens.

The aggregative model is based on a contract among people with different interests; institutions have to make them compatible through on-going mediation work. By contrast, the integrative model implies that goods-sharing makes people happy; institutions must discover and protect these commons; thus, they are interpreters and tutors of commons.

The aggregative market-like model of democracy entails free exchanges among actors to their mutual advantage. The key concept is individual interest. The integrative model of democracy resembles a religious community; there are faith, believers and followers, a mission, and ministers as interpreters of the common mission. The key concept is the 'context of shared social values' (March and Olsen 1989, 118).

The problem raised by the aggregative model is the same as that posed by the market: is it possible to clearly understand and to easily exchange needs and values seen through the lens of interests? In other words, is it possible to aggregate individual preferences into one converging policy solely with the instrument of exchange? Moreover, according to March and Olsen, there is a second problem: an exchange may be highly unbalanced if the initial actors' resources are unevenly distributed.

The key problem in the integrative model is the moral integrity of delegates in the institutions. These delegates are constantly tempted to pursue their own interests rather than to work for the common good. They do this, not because they are corrupt, but because they tend to mix their own political affirmation with the mission of the institution. They perceive themselves as completely devoted to the cause, risking to identify their own individuality with the institution they lead. A second problem is competence. How can the institutions ensure equal and competent participation in the definition of common goods? Numerous factors may distort this process: educational level and rhetorical capacity may affect a citizen's ability to influence the formation of 'commons'.

In general, the shortcomings of the March and Olsen dual model (see Sjöblom 1993) concern the same problem that Hirschman (1970) encountered when he initially framed his analysis with the dichotomy between exit and voice. He was unable to justify the capacity of political systems to remain united because both voice and exit are protest against the system rather than 'glue' for it. For this reason, Hirschman introduced a third case – loyalty – which was not opposite to the two others but was able to justify the long-term support for a political system beyond the occasional experiences of voice and exit. The life of a public institution is mostly characterised by low-degree continuous loyalty, while moments of voice or exit are sharp and infrequent. A common 'exit' act is not to vote during general elections, while a 'voice' act is to leave the ballot paper blank. There are quite a few other moments when voice or exit can be expressed, and this is a major problem of representative democracy.

For this reason, a triadic model must be considered a more appropriate representation of political institutions even at local level (Sharp 1984). To prefer a triadic model does not mean to reject the intuitions of the aggregation and integration patterns; on the contrary, an inspiring operation concerns how the dichotomy overlaps with triadic models. There is a partial overlap with Ouchi's (1980) version of the triadic model: *market, hierarchy, and clan*, which is a development of Williamson's (1975) theory of transaction costs reduction (which is a dualistic pattern). The aggregative model is similar to the ideal type 'market', while the integrative one resembles the 'clan': a group united around certain strong values, according to an explicit reference by Ouchi to Durkheim. In this way the ideal type 'hierarchy' remains outside the March and Olsen's model, raising questions as to how institutions can work without subordination relationships.

In other cases – Polanyi's (1977) version for example – the overlap is even more problematic, with the sole exception of 'market exchange', which is again similar to aggregation of interests.[1] The triadic models in some way split the integrative pole into two, distinguishing between a more rigid top-down model of democracy, where an elite can easily determine the *volontà popolare* ('redistribution' according to Polanyi), and a more flexible deliberative democracy paying closer attention to frequent, competent and equal listening to the people. The latter is a sort of generalised 'reciprocity' according to Polanyi, even if it is not clear when and how binding decisions are taken in such a virtuous exchange.

In any case, the introduction of a third pole seems to give rise to overly static and formal democracy models. For example, Powell (1990) calls the third pole a 'network', which means a constant and trustful exchange of information and goods that gives substance to how people represent their will in the public arena. Furthermore, participation has a material side that greatly affects the chances of being included in a decisional arena. Thus, participating requires, on one side, time, money and competence (Brady et al. 1995), on the other inclusion in

1 In Polanyian triadic types of exchange – reciprocity, redistribution and market exchange – the last type is very similar to aggregation because political resources (votes, ideology, patron/client relations, etc.) are continuously negotiated by parties representing clear interests.

networks which offer chances to be rightly informed and mobilised. Otherwise, political participation is a very formal right, a sort of passive attendance at one place-moment (the election day). Thus, even the fairest design of participation can be emptied if the material (income) and social (connection) conditions of people are not assured.[2]

The three material conditions mentioned – time, money and competence – are strictly linked to the kind of occupation. Once the term 'social class', a precise position in production relationships, would have been used; but today the term 'occupation' better captures the material and immaterial resources of a person who wants to participate in public affairs. According to the kind of occupation, it is possible to guarantee the basic personal conditions for participation: for example, a sufficiently high income to allow extra work time, or a job that favours moments of learning and involvement, a sort of 'gymnasium' for public moments of participation.

The third pole (network) highlights the dynamic and concrete condition for the involvement of individual persons, groups and institutions. Effective participation comes about through informal networks behind the formal facade of public meetings and councils (Diani 2004). Effective participation is guaranteed not only by democratic institutions (Etzioni 1964) but also by relational resources which provide valuable information and material support for candidates. To be considered that all these chances coming from network are analytical dimensions. Concrete networks can be very elitist and asymmetrical either in their inside or in their outside relationships (Hadjimichalis and Hudson 2006).

Italian Local Action Groups

One way to evaluate the usefulness of different theoretical models is to consider an empirical case. In the following such a test is made using Italian local action groups (LAGs). In a very synthetic scheme, Italian LAGs can be represented as a position point on a continuum between aggregation and integration (Figure 9.1). Italian local action groups, represented by the star in the Figure 9.1, are located towards the aggregation pole by March and Olsen. The reason for such location is that local action groups are essentially means to aggregate local interests, in order to receive a quota of public funds, under the umbrella of a generic 'rural' label. The decision-making process is based on 'bargaining' according to the ideal types developed in Chapter two in this book (see Table 2.1). There is weak integration around the value of rural as a 'territory', conceived as a delimited space that contains

2 In Italy, assurance of the material conditions for political participation is (in 2013) at the centre of heated debate on the 'costs of politics'. One part of public opinion considers the public grants to political parties, and to people elected, as immoral, while the other part (usually the historical left) sees the public money for politicians as a way to ensure the material conditions for participation by people on low incomes.

Figure 9.1 Usual Position of Italian Local Action Groups on a Continuum between Aggregation and Integration Ideal Types

indivisible goods and needs a unitary promoting institution.[3] Italian local action groups tend towards the aggregative pole because they are essentially mediators between centre and periphery, agricultural and non-agricultural interests, small and big firms, public and private bodies. This 'link in a chain' role was noted for the first generation of European Initiative LEADERs (Osti 2000) and seems to be still the same for later ones (Celano 2005).

Territory is the integrative value, but it is rather weak because it is conceived as a pure marketing operation. Local action groups are better able than a single company to promote an area's image as beautiful, healthy, friendly, etc. However, they are no more than marketing agencies. According to the theory of goods (Ostrom 2005), they should be especially efficacious in developing *club goods*[4] – that is, indivisible goods whose benefits are appropriable only by the members of the territorial community.

The best expressions of this capacity are those Italian local action groups that have been able to become multi-polar agencies; they are more than purely temporary aggregations of interests. As quite stable, autonomous, and multi-task local institutions, they move the star in the figure below towards the centre of the aggregation-integration continuum:

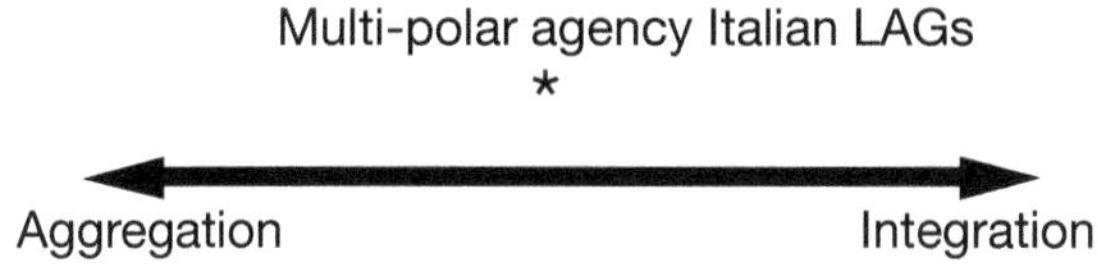

Figure 9.2 Position of Multipolar Agency – Type of Italian Local Action Groups on a Continuum between Aggregation and Integration Ideal Types

3 The term 'territory' has been preferred to 'place' (see Gieryn 2000) because it entails a highly dynamic dimension: place is a space clearly bounded and emotionally lived, territory implies an active, organised and internal action of promotion (see Italian geographers in Pollice 2003).

4 Technically, club goods are without rival consumption and easily excludable. Club goods often include local or tacit knowledge.

In the best cases, local action groups are recognised as somewhat more than temporary aggregations of interests[5], they have the capacity to transform individual interests into more unitary projects usually based on economic development of land-based goods (food, wine, environment, landscape, local heritage, etc.). A single company or economic sector is unable to produce locally-based products or club goods (for example, wine with Protected Designation of Origin or Protected Geographical Indication). They are also unable to aggregate firms in the task of controlling free riders. For these reasons, the most aware companies put a local action group in charge of controlling common goods. The local action group is invested with new authority: to certify the proper mode of production, to rebuke for nonstandard products, and promoting, generally by a mark, the products and the companies that follow the protocols.

Confirmation of this analysis is provided by recent research. The local action groups of the Emilia-Romagna region are considered by the LEADER National monitor task force (*Rete Rurale Nazionale*, RRN) to be those with the most *political* and *functional* autonomy;[6] they have become in our terminology 'local institutions' whose *super partes* role is unanimously recognised. From the LEADER perspective, the Emilia-Romagna Region has given the local action groups the faculty to choose either the eligible municipalities where to intervene as the socio-economic fields of intervention (RRN calls it *political or decisional autonomy*) while reducing to the minimum its own formal control over LAG actions (RRN calls it *functional or administrative autonomy*). This wide double autonomy stimulates the local action groups' capacity for self-evaluation (Secco et al. 2011), creating a virtuous circle between autonomy and self-reliance.[7] Furthermore, the capacity to combine the political and functional autonomy gives institutional prestige to LAG because it is recognised as competent by both the centre and the periphery; by the Region Administration and by local stakeholders.

Budget auditing is an important example. Local action groups often complain about the abstractness and uselessness of expenditure accounting to the Region and then to the EU. This operation is not a chance to learn how to spend money but only a bureaucratic task. Instead, budget formation and evaluation should be

5 On temporary organisations, especially their role in local development, see Sjöblom et al. (2012).

6 An important theoretical reference for the political and functional autonomy of local bodies is the model developed by Page (1991), who distinguishes between the legal and political status of local governments. The former concerns the formal competences given to the local bodies, the latter concerns the capacity of such bodies to interact with central government. According to Page, local bodies in Southern Europe have lower legal status and higher political status because of their capacity to intertwine in particularist manner with individual sectors of the national government. For centre-periphery relationships in LEADER, see Rizzo (2009).

7 Self-monitoring is a basic assumption of local development. Cognitive autonomy means the capacity to know the needs, instruments and relationships more favourable to endogenous development.

Figure 9.3 Number of Local Action Groups in Each Italian Region's 'Rural Development Programme' (2007–2013)

operations useful for understanding not only the local action group's internal skills but also the resources and capacities of the territory where local action group seeks to promote development.

Among the Italian regions, the political system of Emilia-Romagna can be considered, generally speaking, as an example of the integrative model 2 by the March and Olsen theory. Emilia-Romagna's integrative style is distinct from the more aggregative style of the nearby Veneto region. That seems true not only for LEADER but also for other policies (see Messina 2001a and 2001b). For example, the Emilia Romagna Region has created special centres for industrial innovation, while in Veneto this initiative, so crucial for local development, has been left to spontaneous action by professional associations.

Further confirmation is provided by a general picture of rural governance in the two regions:

> Veneto and Emilia-Romagna (see Figure 1) are among the richest Regions in Italy and represent the so-called 'third Italy', whose economic development was based on the interlinking of small agricultural and industrial enterprises organized in specialized districts, in the 1970s. The two regions differ, however, in their rural policy governance system: Veneto has a traditional 'mixed' system of governance in which most decisions are made at regional level and only some aspects of delivery are devolved, while Emilia-Romagna has set up a fully decentralized system of governance in which the Provinces have a more significant role in all the different stages of policy design and implementation. (OECD 2009, 108)

Despite the large evidence on LEADER, the picture of its relation to local democracy is rather poor. The integrative-aggregative dualism only vaguely helps to answer the question of local democracy. The supposed integrative style of some Regions or local action groups assures more involvement only at the symbolic level (people feel themselves to be united or identified to a precise place), but the quality of participation remains obscure. Even if there is a reference to a bottom-up or deliberative method (see Chapter 2, Table 2.1), it is well-known that this takes place only indirectly through participation via two channels: the representatives of municipalities and those of professional organisations (e.g. farmers' unions) in the LAG council.

According to our premises, a triadic model should yield better understanding of the democratic potential and effectiveness of the LEADER approach. We need to shift from a linear 'aggregation-integration' model to a triangular one comprising networks, hierarchy and market.

There are numerous, even if weak, signals of the evolution of LEADER organisations towards networks. In fact, Italian local action groups often have:

- large memberships without a rigid prominence of one member over others
- the criterion of one vote for each member of the board
- some non-governmental organizations included in the board
- legal status of an association or cooperative.[8]

8 In 2010, thirteen local action groups were formed in the Sardinia Region: four opted for the legal status of foundation of participation (a *de facto* mix of patrimonial foundation and involvement of individuals), eight as recognised associations, and one as a consortium. See: http://www.galmarghine.it/home/notizie-e-aggiornamenti/100-analisi-del-processo-partenariale-e-dellistituzione-dei-gal-della-sardegna.html.

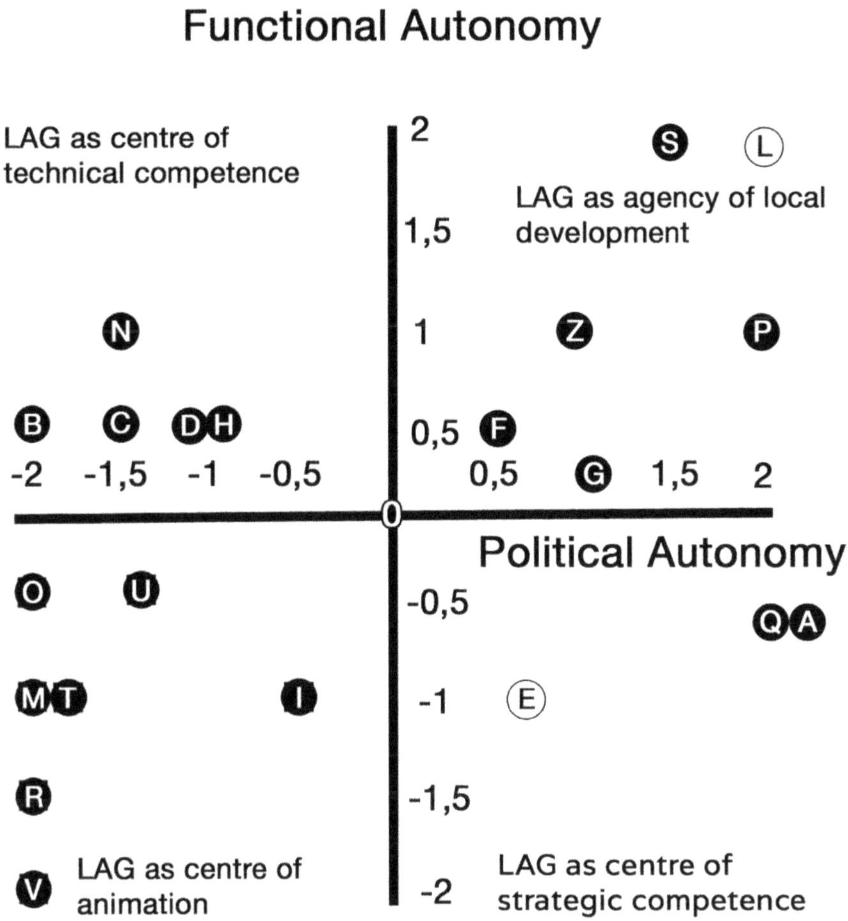

Figure 9.4 Classification of LEADER Regional Rural Programmes* according to the Political and Functional Autonomy of LAGs

Notes:
* LEADER approach is incorporated in the regional rural programme or plan
L = Emilia Romagna; E = Veneto; the other letters indicate the other Italian regions
Source: D. Cacace, et al. 2011.

Since the origins of LEADER program, the local action groups insist on partnership, bottom-up procedures and co-financed projects; these are the ingredients of a 'network' whose distinctive feature is reciprocity. The tendency is thus towards the network pole. The introduction of a triadic model and of network/reciprocity-type regulation entails changes in the vision of local action groups' local democracy potential. It provides a criterion of analysis, which is broader than that deriving from the integration/aggregation dichotomy; that is, participation is seen in terms of inclusive relationships, not just as the right to vote at any time. It consents to look for signs of reciprocity among the LAG partners, beyond formal relationships and distribution of roles. Field research would show sharp differences between the formal model of inclusion in the board and the real decision-making process. A survey on four Italian LEADERs, using network analysis, showed that the board members did not always have the relational centrality assigned by their role (Cimiotti 2006; Osti 2006). For example, in some cases the LAG president was not the most central person in Moreno's sociogram. In others, there was a weak if not conflicting relationship between the LAG board and the staff.

Nevertheless, the most important aspect of introducing a triadic model, when researching the democratic possibilities of LEADER, is the idea of looking at the extent to which LAG's members and investors share the property of enterprises. In other words, how diffuse are forms of economic democracy among the organisations involved in the LAG board or supported by the LAG projects. How far shared are different kinds of capitals (land, technologies and money) among the protagonists of the LEADER endeavour. The assumption is that full and effective participation in an organisation happens when there is a frequent, fair and competent voice and when members at least partially share the economic burden of that organisation.

According to Robert Dahl (1986), the idea behind economic democracy was first to transform workers into shareholders. Later ideas were to deliver stock options for enterprise managers and in Germany to develop the so called dual governance system for stock corporations (the Management Board and the Supervisory Board). It has been an underlying ideal of western societies, which was never completely erased by industrial relationships system. Moreover, economic democracy is one of the above-mentioned concrete resources for participation: without a job and a capacity to accumulate in order to become a shareholder, it is difficult to participate in public life.

One wonders whether local action groups are totally outside these forms of involvement into the companies' property and management. There are reasons for thinking that local action groups are not so distant from a certain kind of economic democracy. We have to move from classical political participation (voting and deliberative democracy, e.g. consensus conferences, citizens' juries and so on) to a broader vision that includes democracy within the economic field. In Italy there are at least four fields with ongoing experiments of this type:

- Producer cooperatives (Italian labour and production coops); all cooperatives must comply with the 'one head one vote' principle and the profits distribution constraint (Borzaga and Defourny 2001).
- Public or diffuse shareholding companies; it is true that often a shareholder minority or the manager can easily control the company; but in some cases, the Italian popular banks of cooperative type can be examples of economic democracy (Boscia and Di Salvo 2009).

- Pacts/alliances between producers and consumers (solidarity purchase groups); there are now hundreds of these organisations in Italy, even if they are rather small. They are formed by a vanguard of very responsible consumers (Brunori, Rossi and Guidi 2012).
- Divisional organisation structure of the company; of course, only large companies can be organised in terms of a network of relatively autonomous units; in Italy it is easier to find networks of firms, a formula now officially promoted by the state (Ferrari 2010). A firms' network has its own legal status distinct from that of the single member firms and is built around an entrepreneurial project, which is evaluated and possibly financed by the Ministry of Industry.

Which of these partial forms of economic democracy are present in the Italian LEADER? This is a matter for research not yet undertaken but certainly applicable to local action groups and to the co-financed companies. To date, the quality of the economic partnerships stimulated by the Local Action Plan has not been evaluated: usually, the evaluation criteria regard three aspects: compliance with the rules (spending according to the timetable), impact on job creation, and involvement of different actors in the decision-making process. The quality of economic enterprises, both for-profit and non-profit, is rarely analysed. The focus may be on environmental aspects; thus, the involvement of organic farms or firms adopting eco-friendly devices is positively evaluated. A certain attention is also paid to workers' rights, highlighting the presence of labour relationship certification (SA8000). The presence of firms involved in corporate social responsibility projects is sporadic.

Nonetheless, constant undervaluation of third-sector actors in the LAG-financed projects is evident in Italy. It bears out the general impression that LEADER has been an experiment in local democracy circumscribed to local authorities and the local branches of vested interests. Once such interests pivoted around primary sector, but nowadays they have been progressively enlarged to include the tourism and crafts sectors. The inclusion of social cooperatives[9] – non-profit enterprises with collective goals – appears to have been limited to a few pioneering cases. This also depends on the weakness of the third sector. Unfortunately, there are no studies on the more or less active role of cooperatives and volunteer associations in the local action groups and local action projects.

An even more interesting research field concerns the 'stability' and 'openness' of local economic networks included in the local action groups. The expectation is that their stability is quite high because of the good tradition of the Italian industrial districts; in many places local action groups have simply overlapped with an existing robust socio-economic structure, a sort of local corporatist system

9 Social cooperatives are in Italy recognised by the law. Their aim is to work *with* or *for* disadvantage people. The former (*with*) are workers' cooperatives, the latter (*for*) are providers of social services.

(Hernes and Selvik 1981); in others, especially in south Italy, local action groups have been relatively successful *first cases* of local institutions/firms cooperation.

The evaluation is more severe as regards the openness of local networks. According to Barca (2011), this is the greatest challenge for Italian's as well as European Union's local development policies. To date, compassionate redistribution and communitarian self-reliance have been the models pursued. Both of them lack a strong drive towards exogenous institutions, especially towards ones with a high content of knowledge and expertise (universities, research centres, technological poles, firms incubators); both models are at risk of being captured by rent-seeking local elites; these elites do not belong to the traditional land rentiers of rural economy (Marini and Mooney 2006); instead, they are shaped by local politicians able to keep their power thanks to good relations with small local firms and entrepreneurs organisations (Cingano and Pinotti 2009). That is another way for reaffirming the presence of local corporatism. This is a pattern of local and economic democracy which is not the most open to internal and external new social forces.[10]

Conclusions

Overall, it seems that the integrative/aggregative model is too narrow and past-oriented: that is, related to state/market compromises or liberal democratic systems, even if at local level. It does not include the grassroots participation needs that have emerged in single spots (Porto Alegre, Seattle, New York's Zuccoti Park), but that represent a world movement and sentiment. If we apply the old dual model, we view local action groups as a middle point between public and private, and as a modest compromise between local economic interests and the common value of the territory.

If we, instead, use a triadic model we can measure the degree of democracy *inside* economic actors and economic exchanges, e.g. within labour relationships and consumers/producers relationships. Labour relationships are rarely analysed, because LAGs' projects usually involve only small firms with a dominant family structure and few employees. Consumers/producers relationships are seen in terms of client 'satisfaction' (marketing), while the political dimension of consumerism being almost entirely neglected; 'vote with your wallet', we would say, and from a LEADER perspective, 'orient the project according to the ethical values of consumers'. Cases of this type are quite rare in the Italian local action groups, but at least the triadic model allows them to be seen in potential terms.

Moreover, the emphasis on networks, reciprocity and trust relations, helps in investigating a rather neglected aspect of local development policies, of which LEADER is certainly an emblem. The reference is to more general or universal solidarity; local action group is an example of local loyalty: people do not 'vote

10 On chances of local corporatism to be really democratic see Schmitter (1983).

with their feet' and hence do not leave their locality when its life conditions worsen. But, what about more general reciprocity with other areas? There is decentralised cooperation among the European Union's policies; LEADER itself had a special measure for exchanges with other areas. That was certainly a great opportunity to open the borders of rural areas. Nevertheless, such openness is not enough at a time when global problems like food shortages and energy crises are growing; new forms of local democracy, which able to include distant areas have not been explored.

Thus, the final answer to the initial question, summarised in the title of this article, leans toward the idea of weak local aggregations of interests. The majority of Italian local action groups are examples of a mesocorporatist figure with a social base larger than the traditional agriculture chain but still limited to few other economic sectors. Neither tourism nor environmental services – two strong pillars of multifunctional agriculture – have been able to enlarge the financial participation of a broader constituency. The best local action groups have evolved towards development or certification agency. Consumers' groups and social cooperatives have been almost absent from boards as well as from co-financed projects.

The weak financial and social enlargement of Italian local action groups is probably due to an old urban structure of the countryside. This seems contradictory, but in fact rural and urban are so intertwined in Italy that local action (of development) cannot work without involvement of the nearby town. It is the rural-urban link which has shown elements of innovation in recent years: people from both places exchange new goods (savings, quality food, green care, renewable energy sources). Local action groups have probably deployed on the old lines of agro-tourism, still offering typical artisan and agriculture products to urban consumers. This exchange is not enough for the survival of rural economies; new local action groups must enlarge not only the range of products/services, but also the variety of ways and the people with which they are produced.

References

Barca F., 2011. *Alternative Approaches to Development Policy: Intersections and Divergences*. Paris: OECD Regional Outlook.

Borzaga C. and Defourny, J., eds, 2001. *The Emergence of Social Enterprise*. New York: Routledge.

Boscia, V. and Di Salvo R., 2009. The Theory and Experience of Cooperative Banking. In: Boscia, V., Carretta, A. and Schwizer, P., eds. *Cooperative Banking: Innovations and Developments*. Basingstoke and New York: Palgrave MacMillan.

Brady, H.E., Verba, S. and Lehman Schlozman, K., 1995. Beyond Ses: A Resource Model of Political Participation'. *The American Political Science Review*, 89(2), pp. 271–94.

Brunori, G., Rossi, A. and Guidi, F., 2012. On the New Social Relations around and beyond Food. Analysing Consumers' Role and Action in Gruppi di Acquisto Solidale (Solidarity Purchasing Groups). *Sociologia Ruralis*, 52, pp. 1–30.

Cacace, D., Di Napoli, R. and Ricci, C., 2011. *La valutazione dell'approccio Leader nei programmi di sviluppo rurale 2007–2013: un contributo metodologico*. Rete Rurale Nazionale 2007–2013, Ministero delle politiche agricole alimentari e forestali, Roma.

Celano, S., 2005. Governance, democrazia locale e degenerazione delle politiche, Rassegna Italiana di Valutazione, a. IX, n. 33, pp. 77–94.

Cimiotti M.L., 2006. Il Gruppo di Azione Locale come Rete Relazionale. In: Cavazzani, A., Gaudio, G. and Sivini, S., eds. *Politiche, governance e innovazione per le aree rurali, Edizioni Scientifiche Italiane*, Napoli, pp. 271–98.

Cingano, F., Pinotti, P., 2009. *Politicians at work: The private returns and social costs of political connections*. Bank of Italy Temi di Discussione (Working Paper) No. 709. Available at: http://www.bancaditalia.it/pubblicazioni/econo/temidi/td09/td709_09/td_709_09_en/en_tema_709.pdf. Accessed on 5.7.2013.

Dahl, R.A., 1986. *A Preface to Economic Democracy*. Berkeley: University of California Press.

Diani, M., 2004. Networks and Participation. In: Snow, D.A., Soule, S.A. and Kriesi, H., eds. *The Blackwell Companion to Social Movements*. Oxford: Blackwell, pp. 339–59.

Etizioni, A., 1964. *Modern Organizations*. Englewood Cliffs, New Jersey: Prentice-Hall.

Ferrari, C., 2010. The Italian 'Network Contract': a New Tool for the Growth of Enterprises within the Framework of the 'Small Business Act'? *Columbia Journal European Law*, F. 77.

Gieryn, T.F., 2000. A Space for Place in Sociology'. *Annual Review of Sociology*, 26, pp. 463–96.

Hadjimichalis, C. and Hudson, R., 2006. Networks, Regional Development and Democratic Control. *International Journal of Urban and Regional Research*, 30(4), pp. 858–72.

Hernes, G., Selvik, A., 1981. Local corporatism. In: Berger, S.D., ed. *Organizing interests in Western Europe: pluralism, corporatism, and the transformation of politics*. Cambridge: Cambridge University Press, pp. 103–19.

Hirschman, O.A., 1970. *Exit, Voice, and Loyalty*. Cambridge: Harvard University Press.

March, J.G. and Olsen, J.P., 1989. *Rediscovering the Institutions. The Organisational Basis of Politics*. New York: The Free Press.

Marini, M.B. and Mooney, P.H, 2006. Rural economies. In: Clocke P., Marsden T., Mooney P.H., eds. *Handbook of rural studies*. London: Sage. pp. 120–35.

Messina, P., 2001. Due modi di regolazione politica dello sviluppo locale. I casi del Veneto e dell'Emilia-Romagna. *Istituzioni del Federalismo*, 1, gennaio–febbraio, 145–77.

Messina, P., 2001b. *Regolazione politica dello sviluppo locale: Veneto ed Emilia-Romagna a confronto*. Torino: Utet.
OECD, 2009. Policy Assessment. In: OECD. *OECD Rural Policy Reviews*. Italy 2009. Paris: OECD Publishing.
Osti, G., 2000. LEADER and Partnerships: The Case of Italy. *Sociologia Ruralis*, 40(2), pp. 172–80.
Osti, G., 2006. Il significato politico ed economico dell'esperienza LEADER: quattro casi di studio nell'Italia settentrionale. In: Cavazzani, A., Gaudio, G. and Sivini, S., eds. *Politiche, governance e innovazione per le aree rurali, Edizioni Scientifiche Italiane*. Napoli: Edizioni scientifiche italiane, pp. 541–69.
Ostrom, E., 2005. *Understanding Institutional Diversity*. Princeton, New Jersey: Princeton University Press.
Ouchi, W., 1980. A framework for understanding organizational failure. In: Kimberly, J. and Miles, R., eds. *The Organizational Cycle Life*. San Francisco: Jossey Bass, pp. 395–429.
Page, E.C., 1991. *Localism and Centralism in Europe. The Political and Legal Bases of Local Self-Government*. Oxford: Oxford University Press.
Polanyi, K., 1977. *The Livelihood of Man*. New York: Academic Press.
Pollice, F., 2003. *The role of territorial identity in local development processes*, Proceedings of the Conference THE CULTURAL TURN IN GEOGRAPHY, 18–20 September 2003, Gorizia Campus.
Powell, W.W., 1990. Neither market nor hierarchy: network forms of organization. *Research in Organizational Behavior*, 12, pp. 295–336.
Rizzo, F., 2009. Local institutions and agrarian structures matter in LEADER: Case studies from Finland and Italy'. *Finnish Journal of Rural Research and Policy*, Special Issue 2, 17, pp. 71–84.
Schmitter, P.C., 1983. Democratic theory and neo-corporatist practice'. *Social research*, 50(4), pp. 885–928.
Secco, L., Da Re, R., Birolo, L. and Cesaro L., 2011. La qualità della governance in ambito rurale: prime riflessioni sull'auto-valutazione dei GAL LEADER. *Agriregioneuropa*, 7(26).
Sharp, E.B., 1984. 'Exit, Voice, and Loyalty' in the Context of Local Government Problems. *The Western Political Quarterly*, 37(1), pp. 67–83.
Sjöblom, G., 1993. Some Critical Remarks on March and Olsen's Rediscovering Institutions. *Journal of Theoretical Politics*, 5(3), pp. 397–407.
Sjöblom, S., Andersson, K., Marsden, T. and Skerratt, S., eds, 2012. *Sustainability and Short-term Policies. Improving Governance in Spatial Policy Interventions*. Aldershot: Ashgate.
Williamson, O.E., 1975. *Market and Hierarchies: Analysis and Antitrust Implications*. New York: Free Press.

Chapter 10
Bottom-up Initiatives and Competing Interests in Transylvania

Dénes Kiss and Enikő Veress

Introduction

As a consequence of European Union membership, new methods of governance appeared in Romania. The endogenous approach to socio-economic rural development was a novelty for Romania, in which a state-led, exogenous socio-economic developmental approach had been dominant throughout the twentieth century and the post-communist period. During the negotiation and after the accession of Romania to the EU in 2007 the new model of endogenous, territorial-based socio-economic development has been implemented step by step.

LEADER was implemented 20 years earlier in the EU. Even though there have been critics of the way the programme worked, it has been widely appreciated as one of the successful programmes of the EU (Ray 2000). The success of the programme has led to the inclusion of the programme as one of the pillars of rural development policy.

Since the programme was included in the Romanian Rural Development Policy, there have been doubts as to how workable it is in Romania as it is based on bottom-up initiatives of local actors and the mobilisation of local, endogenous resources – factors that were lacking in Romanian rural society or at least have been very inactive in recent history. The application of the LEADER programme in Romania is therefore worth analysing.

While elaborating this chapter, the implementation of LEADER is taking place. Therefore, this process is analysed using a pilot study. First, we briefly outline the socio-economic context in which the LEADER approach has been carried out in Romania: the situation of the Romanian countryside, focusing in particular on the various institutional stakeholders, as well as the legal framework that regulates the implementation of LEADER in Romania. And thereafter, in the pilot study, we focus on how the partnership was created for a Local Action Group (LAG).

Our theoretical standpoint is partly based on the questions raised by Mark Shucksmith. Shucksmith (2000) points out that development involving territorial cultural identity might exclude a part of the population if they do not feel an affinity towards the constructed cultural identity. This may happen mostly in ethnically heterogeneous societies where ethnic groups try to build parallel cultural identities. So, it is questionable if LEADER can surmount local conflicts of interest and can integrate those (e.g. ethnic) groups that are underprivileged into the power structure.

Then again, another question arises: if external, exogenous actors are decisive in the implementation process of the LEADER programme, (which is very likely if there is no previous history of endogenous development) do they import the external, macro-societal conflicts of interest in local processes? This is why our analysis centres on the extent to which macro-societal conflicts affect development policy. We focus on conflicts that occur between the political, interest and professional groups, between rural and urban population as well as the conflicts of interest between ethnicities.

The Social and Economic Context of the Romanian LEADER: The Postcommunist Romanian Countryside

The most important developmental processes that define and influence the Romanian countryside have similarities to the other Eastern European countries. Despite such similarities, the Romanian countryside is considerably different from the countryside of most former socialist countries of the region. A deciding factor for the developmental path of rural Romania was that until the mid-twentieth century as well as in the postsocialist period, the countryside relied heavily on agriculture. The modernisation of agriculture has taken place through the establishment and the development of cooperatives and state farms. Due to the transformation of the mid-sixties, 95 per cent of the arable land of the country was intensively cultivated (Hunya 1991).

These agricultural farms were more efficient than they were previously which led to a decrease in agricultural manpower. The surplus labour force was absorbed by the newly-built socialist industrial plants. The former agricultural population either commuted from rural to urban areas or migrated to the emerging urban settlements. As a result of this migration, the marginal areas of countryside were significantly depopulated. The situation was better in suburban villages, from which daily commuting was possible. The rural population decreased slowly and the working part of population decreased much faster (Table 10.1). In 1992, about half of the country's rural population lived in villages and a relatively large part worked in agriculture. The role of an agricultural occupation among the rural population was important because, compared to other socialist countries, industry was strongly concentrated in the cities. For example the small processing industry so common in agricultural collectives in Hungary was practically non-existent in Romania.

When analysing the socialist period, the analysis of the development of public institutions (public administration, education and health among others) is often neglected, even if this has been important both for the rural population's social welfare and for employment opportunities in this sector. Administration was organised by townships and one rural township included on average five villages.

The core village of such a township had a certain poll-position compared with the other villages. As a particularity of Romanian socialist rural policy in the peripheral villages, these villages were left out of implementation and development measures.

The breakdown of socialism has led to a fundamental realignment of rural society. The two most significant changes in the economy were the marginalisation of workers employed in industry located in the countryside, and the reorganisation of agriculture. The socialist industrial units, particularly the cooperatives were closed in a couple of years, lands were redistributed to their former owners (there have been two laws of land redistribution in Romania which have largely contributed to the non-existence of land concentration), and an overwhelming majority of small private plots still characterise Romanian agriculture. The substantial dismissal from urban working places together with a quite arbitrary privatisation of urban cooperatives has led to a recurrence of the traditional rural way of living. The sudden increase of impoverished peasants has been defined as a process of 're-peasantisation' of the rural population, for whom the countryside as well as agriculture in many cases acted as a 'social buffer' (Veres 1999).

As can be seen in Table 10.1, the ratio of the agricultural employment decreased until 1989, but grew during the nineties. The importance of employment in rural non-farming sector decreased after the transition, but remains significant. Employment in the public sector has shown constant growth.

Table 10.1 The Share of Employment in the Different Occupational Sectors in Romanian Rural Areas 1977–2002

	1977	1992	2002
Agricultural sector	62.4	50.3	56.1
Non-agricultural sector	33.9	43.1	35.4
Public sector	3.7	6.6	8.5
Total	100	100	100
Total employed rural population (thousands)	5,506	4,351	3,759
Total rural population	10,749	10,371	10,220

Sources: National Institute of Statistics Romania 2004, Minnesota Population Center and National Institute of Statistics Romania 2010.

After the transition, the long economic crisis condemned the rural population to the countryside, and has precipitated in the first generation of urban dwellers some degree of return to their 'homeland' (Rotariu-Mezei 1999). The rural population grew during the 1990s followed by a massive international out-migration which again decreased the numbers. In 2005, 45.1 per cent of Romania's population was living in rural areas, i.e. approximately 9.7 million people. The ratio of the rural population is roughly the same as 15 years earlier at 45.7 per cent (National Institute of Statistics Romania 2006). Some of the decrease can be partly attributed to the fact that some villages were declared towns, as well as to international out-migration. The average age as well as the percentage of the inactive population has further increased: in 2002, 63.2 per cent of the rural population from Romania was inactive (see Table 10.1).

Institutional Actors in Romanian LEADER

LEADER is based on the local initiatives of different actors from the rural public sector, business sector and civil society. This idea assumes a relatively well-developed economy and a strong civil sector, and presumes that the government is able to cooperate with them. However, in various European countries these institutional actors have progressed to varying degrees, and seemingly similar institutional structures disclose in fact very different organisational cultures.

The most important public organisations in the Romanian countryside are represented by local administration including the mayor's office and educational and healthcare structures. The administrative institutional structure has largely retained its structure from the socialist era. In Romania's regional administration, the lowest unit is the rural municipality, which coincides with the OECD's LAU2 unit. In 2002, the country had 13,042 villages that were concentrated in 2,727 municipalities. On average a rural municipality includes 4.8 villages, with an average population of 3,762 people (National Institute of Statistics Romania 2004). We must mention that in Romania there has been no administrative unit of micro-regions. In most cases the postcommunist administrative structures are the same as stated during the administrative reform in 1968. The units of the local administration consist of the local council that is led by the mayor, deputy mayor and secretary. Their economic power was strengthened by the re-privatisation of communal agricultural lands and forests, which gave incomes in addition to the funds allocated from the central budget. Depending on the size of the population, the staff of the local administration varies from 10 to 30 persons.

Economic enterprises can be either agricultural or non-agricultural. In agriculture, the closure of large-scale socialist industrial units and the subsequent re-peasantisation (Veres 1999) of the rural population has resulted in dually-structured, heavily divided agriculture. Most of the agricultural sector consists of small subsistence farms, which are unable to reach the markets and are working outside of the formal labour market.[1] Besides this segment, there is also a narrow segment of large-scale industrial units, which, although small as a proportion of

1 Out of those occupied in agriculture only 0.2 per cent are employers and 6 per cent are employed, 51.6 per cent are independent producers, 42 per cent are family members (Minnesota Population Center and National Institute of Statistics Romania 2010).

farms, cultivates one-third of the agricultural area of the country (see Table 10.2). The large-scale farms are made up of commercial companies, agricultural units of the local authorities and agricultural companies / corporate farms. Furthermore, there are few cooperatives and associations.

Table 10.2 The Agricultural Units in Romania 2010

Type of unit	Number of Units	Share of units (%)	Cultivated area (%)	Average area (ha)
Individual household	4103404	99.6	65.5	2.2
Legal entity	17843	0.4	34.5	269
Association	1614	0.04	5.30	460
Private commercial	4325	0.10	12.4	398
State commercial	238	0.01	0.40	252
Local authority's	4750	0.12	15.30	447
Cooperative	89	0.002	0.00	37
Other form	6827	0.17	1.10	23
Total	4121000	100	100	3.4

Source: National Rural Development Plan 2010.

The proportion of those occupied in the non-agricultural sector has decreased between 1992 and 2002 from 43.1 per cent to 35.4 per cent. This can be explained by the decrease in the commuting possibilities towards the urban area. The number of commuters decreased by half in the 1990s from 1,200,000 to 600,000 (Sandu 2005). Despite slow growth in the number of non-agricultural enterprises recently, their total number is low. More than half of these companies are retail outlets. In addition, processing, transportation, construction and rural tourism have a share of the rural non-agricultural economy.

The civil sector is weakly developed. Even if non-governmental organizations (NGOs) have emerged in the villages since the nineteenth century (mostly in the western part of the country), this sector was torn apart by the communist regime when all civil initiatives that were not politically controlled were completely banned. Thus present-day Romanian civil society is smaller in size than the European average or even the Eastern European average (Salamon et al. 2003). There is a significant inequality in the territorial dispersion of NGOs since they are better represented in the Western part of the country (Andersen et al. 2010). There is also a disparity in ethnic dimensions. The ethnic Hungarians are the largest ethnic minority of the country and in Transylvania their organisational activity is significantly higher than that of the other ethnic groups (Kiss 2010).

During the search for the potential implementers of the LEADER programme, we can conclude that local authorities are the most stable and territorially most evenly-situated rural organisations. The number of small and medium-size enterprises is low; economic activity is mostly based on more or less informal family enterprises and craftsmen. The latter category includes the traditional crafts. Because of the splitting-up of the agricultural sector the potential partners in LEADER projects are the large commercial farms, which hardly represent the interests of small landholders. The non-profit organisations constitute major partnerships in the country's Western region, Transylvania, especially in areas populated by ethnic Hungarians. All three organisational types strongly concentrate on the municipality centres, which leads to an underrepresentation of the interests of more peripheral villages.

The Rural Policy and the LEADER in Romania

Until the accession of Romania to the European Union one could hardly speak of a rural development policy. The transformation of the agricultural policy into a more territorial-based policy has been the result of the accession in 2007 of Romania into the EU. It was this moment that the National Strategic Plan for Rural Development (NSPRD) for 2007–2013 has been elaborated, followed by the National Plan for Rural Development (PNDR). The latter – reflecting the EU rural development plans – is made up of four pillars, the LEADER stands for the Fourth Pillar. According to the plans, the implementation of the programme in Romania is done through three measures. One of the measures supports the fruition of the real rural development projects of the LAGs throughout the LEADER project. Another measure supports the establishment of partnerships among the LAGs. The third measure (measure 431) is meant to help with the establishment of the LAGs, and help their sustainable functioning through the acquisition of competences and territorial animation. The implementation of the programme actually starts with this third measure that has two sub-measures. The aim of the first sub-measure (431.1) is to support the elaboration of the strategic developmental plans, the formation of the partnerships that are necessary for the establishment of a LAG. The second sub-measure (431.2) is meant to support the functioning of the LAGs which have already been approved (operation costs, animation and development of competencies).

The 431.1 measure was meant to support the formation of the LAGs, the measure being justified by the acute lack of LEADER-type organisations, and their lack of financial resources necessary to elaborate developmental strategies. This preparatory phase has been divided into a further three phases. The first phase,

'Growing the awareness of the local actors regarding of the LEADER approach' has the aim of supporting the spread of information regarding the LEADER programme. The aim of the second phase was the organisation of special training for the elaboration of development strategies ('Training of the representatives of the potential LAGs'). Subjects of this training come from the participants of the training from the first phase, if they could prove that they are representatives of potential groups that are composed of at least two private and one public institution. The third stage ('Financial support for the preparation of the LAGs' selection') was the last phase before pre-selection, when the potential LAGs have been granted aid in order to be able to make up the dossier needed for participating in the tender process. This support could be applied for by projects elaborated by the partners of the LAGs which had to include the aims, the objectives, the period of implementation, the time needed for the elaboration of the strategy and the budget for the assembling of the complete tender dossier.

Criteria of Eligibility and Final Selection

Understandably the programme could only include territories with rural status. In Romania, settlements that have rural status are the rural municipalities (townships) and the subordinate villages. Besides these rural municipalities, it was possible to include 206 small towns, all of them with less than 20,000 inhabitants and a low degree of urbanisation. In order to prevent the prevalence of the urban interest, however, in a LAG only one small town could participate.

Besides the small towns, a LAG can also include as partners urban-based organisations, if their activity is carried out in the rural area or is concentrated on rural issues (for instance county councils, prefectures, local councils, training institutions, consultancy firms, also rural-based firms). The conditioning tries to limit the prevalence of the interests of these urban-based organisations through the fixation of their maximum share.

The LEADER's Implementation Schedule

The process of formation of the LAGs was considered to have three subsequent stages and as a result, in September 2009, 40 LAGs would have been selected. These LAGs would have functioned as leading models for 40 more potential LAGs. But the real situation could not fully follow the plan. As happened in the first two phases, the planned two-wave selection was overlapped; finally only one selection was made, also with a significant delay, on the 9th of February 2011. Because of various problems during the accumulation of the eligibility reports this belated selection procedure has also been re-examined, and the disclosure of the results was again adjourned sine die (MARD 2011).

Some Preliminaries of the LEADER

For a better understanding of its Romanian implementation we have to mention a few processes prior to the elaboration of the legal framework of the LEADER programme. Such an antecedent process was the creation of the Inter-communal Development Associations. These were institutions with a specific status that have come into being through the association of local authorities representing an intermediate form between public institutions and NGOs. Even though they are partly considered as being organisations of civil society, their functioning is regulated by a separate law (Pop et al. 2007). The aim of their establishment is to accomplish micro-regional development plans and to provide common public services, so their formation is not necessarily connected to the LEADER. In the present LAGs they have partnered as private organisations.

Another premise of the LEADER programme is also related to public authorities and civil society. This is the appearance of a 'local development agent' (also called 'community mediator'), which is included in the public administration system. The idea of training such developmental agents arose from civil society, on behalf of NGOs that focus on regional and rural development. They had faced many problems during their collaboration with the local authorities. They located these problems in the organisational culture of the local authorities, hardly compatible with the 'world of the projects'. For this reason some organisations have started to run projects that aimed to train persons able to bridge these discrepancies. In order to convince the local authorities to employ them, in the last stage of this programme, the trainer NGOs temporarily financed these local development agents employed by the local municipalities. At the peak of the project, some 250 agents were employed. Their wages were ensured for half a year and after that the local authorities could decide if they would keep them on further. In the north-western region most of them stayed on in the employment of the local authorities, playing an important role in the elaboration of partnerships and strategies, as well as the tendering of the mayor's hall for several projects.

The Formation of a Romanian Local Action Group

In the following we will present the formation of a Local Action Group (LAG), based on our pilot study. As the selection of the established LAGs was not performed until our fieldwork, we are able to provide here a description and an analysis of the process of the formation of the LAG. In this process local initiatives and the top-down actions are strongly interwoven. The size and composition of the final LAG, as well as the partnership in formation has undergone several significant changes, so the final territory and action group bears little resemblance to the planned one. During the fieldwork semi-structured interviews and document analyses were used. In the interviews with stakeholders of the implementation process, we asked about their personal involvement and asked them to reflect on the whole process.

In addition, we conducted a document analysis of the complete archive of the first and second LAGs and the online documentation of the third LAG.

The Field and the Initiator-Village

The analysed LAG is situated in the Transylvanian macro-region of Romania, in the proximity of one of the largest cities of the region. It is situated in a territory with a multi-fold relief, where there is less arable land. It is rather hilly and mountainous so it is more suitable for animal husbandry and non-agricultural economic activities. As a result, the local population has been involved for several centuries in performing traditional handicrafts, especially those based on wood. This has resulted in an agricultural population lower than the national average and a larger enterprise density. This is the case especially in the settlements that are closer to the city, while in the marginal villages there are fewer enterprises. Instead in the latter, leisure tourism has developed over the last decade. The ethnic structure of the territory is mixed, besides the Romanian majority there is significant Hungarian and Roma population. During the first half of the twentieth century there were several ethnic conflicts between the Romanians and Hungarians, leading to a latent tension between the two ethnic groups. Even if there has been no open conflict since World War II, ethnicity has a strong influence on the articulation of the local interest groups and conflicts; these are also influenced by the role of ethnicity in national-level politics.

The initiative to establish the LAG studied started in the example village with 1050 inhabitants. The village is the seat of the local government for the municipality, which consists of seven villages. The village is inhabited mostly by ethnic Hungarians (90 per cent), but there is also a smaller Romanian and Roma community. The village has several enterprises in the wood industry, and even a cooperative for marketing the production. Besides numerous enterprises in wood industry, in the last 20 years a successful business in rural tourism has been built up and related to tourism the catering industry has also been developed. The village also has a rich cultural and community life, a village chorus, a drama group, winds orchestra and folkdance group. The village has an important network of non-profit associations, and the community-based activities are almost all led by local actors. This know-how has been useful for mobilising the local community.

The Initiator and the Establishment of Local Partnership

During the establishment of the LAG, at least in its first phase, the central role was played by a local agent for development. The 55-year-old man worked until the fall of communism as a commuter engineer in a factory in the nearby city. He took over important tasks in the life of the local community from 1997–1998 in a project funded by the Soros Foundation. He then worked with information technology in the mayor's hall of the village, taking care of the installation of PCs and IT training for personnel. Later in 1999, thanks to the same fund, the office was connected to the internet. In 2000, he played a major role in the introduction of an innovation in the village: the establishment of a Tele-cottage. To accomplish this project, major support was needed from the local community. As a result, an NGO was established to deal primarily with the running of the Tele-cottage, but also to carry out other local development projects, for example to organise training sessions for different local groups. The funding of these programmes was done through different tenders.

After his involvement in the above mentioned community-development projects, he got involved in a training programme for local development agents. During this, he started to work with the local government, and later he became an employee of it. He also started to develop a more extended network with the group of the local development agents from Romania. Since then he has worked as a tender writer for the local government, becoming a typical representative of the 'project class' (Kovách and Kucerova 2006).

He got connected to the LEADER programme in 2006. In this year, the Romanian Ministry of Agriculture arranged training for potential LAG organisers, who in time would become able to establish partnerships and to establish LAGs. He applied and was accepted onto the training course as a representative of an existing partnership. The partnership relation was formalised with signatures of three local authorities, two NGOs and one business organisation. The partners accepted the responsibility of covering the costs of the training, and some later costs. The local development agent assigned himself to be the one who would complete the tender dossier in case the LAG was selected. Thus, during this first period of establishing LEADER, the local development agent got an official qualification for the formation of a LAG, and the local government created a LEADER office for him.

The Second Partnership

During the first LEADER training, it became clear that the territory covered by the three associated local governments was too small to make up a LAG. So the next task was the further territorial expansion of the partnership relations, first of all by the inclusion of further municipalities. This expansion luckily coincided with a tender won by the NGO of the Tele-cottage, subsidised by a foundation from Hungary.[2] The 4000 Euro support was obtained for the establishment of a micro-region that could be further extended as a LEADER-region. In the project there have been several meetings, and as a result a partnership of eight municipalities came into being. The main driving force in the accomplishment of this project, besides the local development agent, was an NGO from the nearby city, with a role to push forward rural development. They gave significant help in the writing of the tender and provided the necessary trained personnel and know-how. The task of the local agent is to find participants for the project. During the meetings of that

2 The Hungarian foundation works specially for ethnic Hungarians outside Hungary.

project they started to come up with a strategy for the development of the territory covered, respectively a website of the area was developed.

From an organisational point of view, the above programme consists of three major events: two meetings and a field trip, when the participant villages were visited, and local meetings were organised to attract further potential partners. The placement of the first two meetings was in the initiator municipality's two villages. Here, the delegates of the eight local governments and of a few local enterprises and NGOs met. The first meeting was rather an informative one, when the LEADER programme was introduced (which was mostly unknown in Romania at that time), during the second meeting they started to elaborate the first developmental strategy. At the end of these meetings the partnership came together, which hopefully will be able to make up successful tenders as LAGs. Looking at the composition of the partnership (Table 10.3), one can see that while the inclusion of the public authorities was successful, the involvement of the business and the NGO sector was less successful: beside the eight local governments they could attract only five enterprises and five NGOs, half of them being from the initiator village. Organisations of ethnic Hungarians are over-represented, seven out of ten being Hungarian – at least as we can conclude analysing the names of the representatives of these organisations.[3]

Table 10.3 The Composition of the Second LAG-Partnership

		Ethnic-based organisations	
	Number of organisations	**Hungarians**	**Romanians**
Local governments	8	1	7
NGOs	5	5	0
Economic (private) organisations	5	2	3

The total territory of the LAG includes a population of 44,426 and an area of 75,912 hectares. For this region, a first development strategy was also produced. The main role in the elaboration of this strategy was played by the experts of the NGO from the city. It is also an expert of this organisation who gathers data from the local governments, then develops a first version of the strategy which was then put on

3 The Romanian and Hungarian first names differ considerably so it is easy to include the representatives of the organisations in an ethnic category, this goes mostly for the NGOs, where the structure of the organisation is very much ethnic-based (Romanians usually work together in the same organisation with Romanians, and vice versa, Hungarians with Hungarians).

the website of the LAG in formation. The LAG also gets a website, elaborated by the local development agent. This website is a wonderful summary of the events of this period of the LAG-formation, of the LEADER programme, the established partnership, the map of the territory, the statistical data of every municipality (economic and social data), and respectively of the strategy developed.

Probably the most important output of the formation of this second phase of partnership network was that a formal partnering of the local authorities has come into being. At the end of this stage, the partners bring together an Association for Intercommunal Development, which has been signed in 2007 as a legal entity. The participants considered that this association could legally represent the LAG.

After this phase, the process hit a dead end and the events came to a halt. The implementation of the LEADER in Romania was suspended. LEADER-connected projects were still carried out, but they were no longer community-based ones. So for instance the above-mentioned local development agent participated in a field-trip in Poland, organised by the county council, where they visited functional LAGs. The created partnership, organised by the above mentioned city-based NGO for rural development was also involved in other programmes. One programme dealing with information was initiated in 2007. In five counties general presentations of the LEADER programme were made and the schedule of Romanian implementation presented.

The Third Partnership: The Mega-LAG

The official start-up of the LEADER radically transformed the LAG analysed. Until 2011, national funding did not exist for LAGs. Even if there were two programmes with national financial support – the first for training and the second for a field-trip in Poland – they were programmes subsidised partially by the organisers, partially from PHARE (Poland and Hungary: Assistance for Restructuring their Economies) resources. The three programmes presented were financed from Hungary, from financial sources that could be obtained exclusively on an ethnic base by organisations of Hungarians from Romania. There were no other sources of funding from the EU for this. Consulting firms, which made tenders for EU funds were not profitable until then. But after allocation of the Romanian funds for the formation of the LAGs, consulting firms had a real business opportunity.

This was also the case of the LAG analysed. Unluckily, the tender for the development of the strategy and the formation of the partnership was delegated to a firm that was not aware of the existing LAG, and even after they realised the situation, they did not take it into consideration in their LAG design. The economic interests prevailed instead of the real objectives of the LEADER programme, the maximisation of economic benefit and ensuring success in this tender proved to be more important than taking into consideration existing local initiatives, partnerships and strategies.

Beside the economic interest of the consulting firms, at this stage national-level political interests also appeared. Under newly emerged political pressure, several local governments – which were involved as partners for the last three years in the second partnership – decided to sign a new partnering declaration. Thus the instigator of the first and second partnerships was faced with the situation that several partners did not want to participate in the process of becoming a LAG anymore, even if they collaborated before on several occasions and signed a declaration of partnership. This was the death of the second partnership.

The establishment of the new, third LAG network was based on the principle of maximising the financial support that could be obtained. They tried to obtain this through the strict consideration of the selection criteria in the tendering process. Funding depended on a scoring system, where a higher score meant more funding. One of these criteria was the LAG's population. According to the criteria, the lower limit was 20,000 persons. For each additional 6,500 inhabitants, one extra point was received. This led the tenderer to form a territory with a much larger population (the upper limit was 150,000). The number of partners was the second criteria for scoring, in this case the lowest number allowed was nine organisations, and above this, each partnering organisation gave plus points. These two criteria led to the tenderers putting together a much larger LAG in this phase. This resulted in the construction of a mega-LAG for the tender which included 29 local governments with a total population of over 100,000 inhabitants. The LAG includes municipalities from two neighbouring counties, including peripheral mountainous areas, as well as areas with meadows and arable land.

One criterion was related to the know-how needed for the activation of the LAG. This could be justified with certificates of participation in the training. Another criterion was that there should be no overlapping between neighbouring LAGs. These latter criteria led the designers of the mega-LAG to include those municipalities from the target territory with which they did not have political relations. As an outcome, the initiating municipality of the second phase partnership network was included in the mega-LAG. In the end, the whole area of the second partnership was now included in the new mega-LAG. But out of the initial NGOs and business organisations only two were included, as well as the Association for Inter-Municipal Development. And finally the new tender was not completed by the leaders of the second partnership, but by the representatives of a sub-urban municipality of the region. The tender was formulated in collaboration with the local government with the same political affiliation as the then sitting central government. This tender concerned the creation of the LAG-partnership, the elaboration of its developmental strategy and the elaboration of the final tender for the official acceptance of the LAG. This mega-LAG got a subsidy of 49,000 Euros, which paid for the consulting firm.

After winning the tender – during the official era of the LEADER programme– the events of the analysed LAG-formation can be divided in two phases. During the first phase, two foreign field-trips were organised, in the second phase, the formation of the final partnership took place, which led to a further restructuring of the territory and partners included.

The first field-trip was made to Hungary, and the second field-trip to several LAGs in Italy and France. This was a kind of reward-trip for the partners and many of them also brought their families with them. After these field-trips the establishment of the final LAG followed, resulting again in a radical restructuring of the territory and partners included. This final restructuring had several causes. On the one hand, those who worked on the final tender realised that during the selection process the mega-LAG would have no real chances of official acceptance as it was not homogenous enough in covering the area. On the other hand, the representatives of some local governments got into partnerships in other LAGs. Because of this, the consulting firm decided to make two tenders for two smaller LAGs.

At this latest stage of the LAG-formation, the composition of the private partners was also significantly changed, especially the business organisations. It is enough to say that reasons for these changes were connected to political and economic circumstances.

After these changes, the tender of the final LAG was finalised. In this process the planning and organisational work was done with a low level of participation by the involved partners. In fact the dossier was completed mostly independently by the consulting firm; many of the partners were only contacted in order to give their agreement. The trustee firm wrote the development strategy and they considered that in this process there was no need for the participation of the partners. Because of this, it happened that the content of the tender dossier did not reach the local governments that were involved as partners, and the president of the LAG himself could not specify all the local municipalities included in the LAG.

The Struggle of Interests during the Creation of the LAG

The case presented in this chapter can be interpreted as a failure, because the partnership was not built in a bottom-up way and the promising results of earlier work were not used. Results from many years of organisational work and a lot of funds used were simply lost. This setback occurred even though in the second phase of partnership several conditions were met for it to become an officially accepted LAG. The events were a joint result of the different interests involved.

If we consider the *political aspect* of the events, we can see that the political interest seemed to dominate over the rural development interests. During the elaboration of the final LAG, the selection of the partners was done after their political affiliation. The tender dossier for the establishment of the LAG was completed by local government, of which the mayor and the majority of the local council belonged to the governing party, even though they were only secondary actors of the former planning period. This local government had been present in the first and the second phase of partnerships and for the government it was embarrassing that they finally had to disappoint their initial partners. However, they could not say no to a political command that came from above. The situation was regarded in a similar way by the representatives of government and those who were disappointed; they did not blame the mayor for this situation, but instead those political actors who reshaped the situation from above and from outside. It is obvious that the political interests overcame the partnering relations based on territoriality and collaboration.

Taken from an *ethnic dimension*, a Hungarian-initiated and strongly ethnic Hungarian interest-based LAG failed to become officially accepted. The establishment of the first and second partnerships was initiated by a village with a mostly Hungarian population, collaborating with an external consultant NGO, which was also led by ethnic Hungarians and most of the funds used in this bottom-up process were activated using ethnic networks. Further on in the first and second partnerships, there was a majority of Hungarian organisations. As a result of the final reshaping of the LAG, the leadership of the LAG went to the local government of a municipality with a Romanian majority and most of the private Hungarian organisations were left out of the partnership. Thus, the interests of the ethnic majority prevailed against the interests of an ethnic minority. It is important to mention that during the events the ethnic dimension was not explicit. This outcome is in a way an 'unintended' consequence of the top-down intervention in the course of these events.

Taken from the *rural-urban dimension* the conclusion is that rural interests face being damaged when in competition with urban ones. Even though an urban-based NGO had an important role in the first stages of the LAG, the local rural actors had much greater control over the foundation process than they had in the final stage. As the initiator possessed the necessary qualifications for mobilisation and strategy-building and the strategy was mostly ready, we can consider that he could have successfully completed the last tenders as well. And they could have had a good chance for success because the area originally established corresponded to all the points in the selection criteria. Despite this, the funds that were eligible for this last stage were steered to an urban-based consulting firm. The head-office of this tendered LAG moved to a municipality which had economic interests strongly inter-woven with urban investment interests. Among the private organisations, the urban-based firms were over-represented. In this respect the regulation of the implementation of the LEADER can be blamed too, as it allows the participation of the urban firms in the partnership. The regulation only stipulates that these firms have to deal in their activity with the countryside conditions, which can be easily justified by most of the firms.

Finally, considering the *economic dimension* of the events, we must say that the possibility of profit-earning through completing tenders has overridden the rationality of the LEADER programme. Even if in this case the second partnership – thanks to the resources that were assigned for this – could have completed its tender dossier with relatively low costs, the tendering had such an important economic contingency that it could not be left out by the economic actors around and in the contest for financial gain the interest-group of the second partnership seemed to be the weaker.

Conclusions

In the case-study presented we can follow how the basic idea of the LEADER programme was shaped and applied to Romanian society and the events related to the creation of a certain LAG, which starts as a bottom-up story, but ends as a top-down one. In the Romanian context this story is an exceptional one, as the high level of social capital and know-how regarding endogenous development of the initiator village here is not so common. But even so these advantages were not a sufficient precondition for their successful participation in the LEADER programme, when external and higher level actors entered the competition for the available financial benefits through the implementation of LEADER. Once these stronger actors entered the scene, the whole course of events took a radical turn, and the final LAG was completely reshaped compared to earlier versions. When analysing the main stakeholders and interest-groups involved in these events, we can see that the local case was shaped by a few conflicts of interest, which are probably the most determinative for Romanian society in general: the political interest against the professional one (of the rural developers), of economic interests against public interest, the ethnic majority's interest against the ethnic minority's interest, and finally the urban interest against the rural one.

References

Andersen, S., Avram, V., Berceanu, D., Lambru, M. and Vamesu, A. 2010. *Romania 2010. Sectorul neguvernamental – profil, tendinte, provocari.* [*Romania 2010. The nongovernemntal sector - profile, trends, challenges*] Bucuresti: Fundaţia pentru Dezvoltarea Societăţii Civile [Online]. Available at: <http://www.fdsc.ro/library/conferinta%20vio%207%20oct/Romania%202010_Sectorul%20neguvernamental1.pdf> [Accessed 9 December 2014].

Hunya, G., 1990. *Románia 1945–1990.* Budapest: Atlantisz-Medvetánc.

Kiss, D., 2007. *Gazdasági-társadalmi folyamatok az erdélyi falvakban.* [*Social and economic processes in the Transilvanian villages*]. Cluj-Napoc: Presa Universitara Clujeana.

Kiss D., 2010. *Romániai magyar nonprofit szervezetek – 2009-2010.* [*Hungarian nonprofit organisations in Romania – 2009 -2010*]. Cluj-Napoca: ISPMN.

Kovách, I. and Kucerová, E., 2006. The Project Class in Central Europe: The Czech and Hungarian Cases'. *Sociologia Ruralis*, 46(1), pp. 3–21.

Ministry of Agriculture and Rural Development Romania, 2010. *Planul National de Dezvoltare Rurala 2007–2013. Versiune consolidata 2010 iunie* [*The National Plan for Rural Development. The consolidate version in June 2010*] [Online]. Available at: <http://old.madr.ro/pages/dezvoltare_rurala/pndr-versiune-iunie2010-romana.pdf> [Accessed 20 October 2013].

———, 2011. *Anunţ privind publicarea Raportului de Evaluare a Planurilor de Dezvoltare Locală Centralizat* [*The official statement of the Romanian Ministry of Agriculture and Rural Development*] [Online]. Available at: <http://www.pndr.ro/uploads/anunt-publicare-raport-evaluare-plan-dezvoltare-locala.pdf> [Accessed 20 October 2013].

Minnesota Population Center and National Institute of Statistics Romania, 2010. Romania Population and Housing Census 1977–2002 from the Integrated Public Use Microdata Series, International: Version 6.0 [Machine-readable database]. Minneapolis: University of Minnesota.

National Institute of Statistics Romania, 2004. *Recensamantul populatiei si a locuintelor 2002* [*Population and Housing Census 2002*] [Online]. Available at: <http://www.insse.ro/cms/files/rpl2002rezgen1/rg2002.htm> [Accessed 20 October 2013].

———, 2007. Anuarul statistic al Romaniei 2006 [Romanian Statistical Yearbook] [Online]. Available at: <http://www.insse.ro/cms/files/Anuar%20arhive/serii%20de%20date/2006/ASR_2006.pdf [Accessed 20 October 2013].

Oláh, S., 2005. Hozzászólás Kiss Dénes erdélyi látleleteihez [Remarks on Kiss Denes's Transilvanian report]. *Hungarian Kisebbség*, 2005/3–4, pp. 94–107.

Pop, D., Stănuş, C., and Suciu, A., 2007. *Asociaţia de dezvoltare intercomunitară.* [*The associations for intercommunal development*]. Cluj-Napoca: CENPO Policy Brief.

Ray, C., 2000. Editorial. The EU LEADER program: rural development laboratory. *Sociologia Ruralis*, 40(2) pp. 163–71.

Rotariu, T. and Mezei E., 1999. Asupra unor aspecte ale migratiei interne recente din Romania. *Sociologie Romaneasca*, 3, pp. 5–37.

Salamon, M., Sokolowski, S. and List, R., 2003. *Global Civil Society. An Overview*. Baltimore: The Johns Hopkins University Institute for Policy Studies, Center for Civil Society Studies.

Sandu, D., 2005. Romania rural-neagricola azi. *Sociologie Romaneasca*, 4.

Shucksmith, Mark (2000) Endogenous Development, Social Capital and Social Inclusion: Perspectivesfrom LEADER in the UK. *Sociologia Ruralis*, 40(2), pp. 208–18.

Veres, E., 1999. De/re-*ţărănizare în* România după 1989. *Sociologie Românească*, 3.

Chapter 11

Can Renewable Energy Contribute to Poverty Reduction? A Case Study on Romafa, a Hungarian LEADER

Ildikó Asztalos Morell

Introduction

After Hungary's accession to the European Union in 2002, LEADER became a key model for rural development, thus providing the tools to local action groups (LAGs) to define the targets for local development (see also Csurgó and Kovách in Chapter 4 of this volume). FA LEADER,[1] founded in 2008, is based on a consortium of 44 resource-poor, small to larger size municipalities in the areas surrounding an industrial city in northeast Hungary. The strategy of the FA LEADER is to strengthen the position of municipalities and the citizens by increasing their energy self-sufficiency and decreasing their dependency on large-scale suppliers of energy in monopoly positions through the utilisation of renewable energy sources. Further, the FA LEADER's energy villages are to utilise 'clean', renewable energy sources, including agricultural waste, warm sources, wind, kitchen waste, manure, water streams and thermo energy from mines for the production of energy and heat. These energy sources are not being utilised at the moment and getting rid of the waste is an expense for the communities. Thirdly, the project aims to develop the communities through the creation of new workplaces for inhabitants with low levels of education and those who are qualified but currently unemployed. Fourthly, FA LEADER aims to increase the quality of life of the inhabitants through cleaning the communities of garbage and improving slum areas where inhabitants had no resources or a previous willingness to engage in such activities. This would also improve the possibilities for developing village tourism.

Romafa is a specific sub-project of FA LEADER targeting marginalised Roma enclaves aiming at 'promoting the energy production of small communities for decreasing their dependency on social benefits and creating a self-sufficient source of income' by the joint utilisation of renewable energy sources, the development of the traditional, hierarchical system of representation, the support of Romani

1 FA LEADER is one of Hungary's regional leader groups. It is a pseudo-name, as are the names used for sub-projects and persons in this chapter in order to keep them anonymous.

traditions, religion, morality, culture, arts, education and the support of self-sufficient production among Romani households.

Three concrete goals were identified targeting Romani communities through the development of 15 municipal and small regional ecological waste collection and processing 'eco-units' for concrete waste according to EU standards:

1. Biogas generators could be operated by deliveries of biomass gathered by resource-poor Romani (and non-Romani), who could either be compensated with cash payments or energy coupons. The compensation model would increase the self-interest, self-respect and autonomous agency of those participating to increase the maximisation of their inputs. Meanwhile, they could promote their respectability as citizens.
2. Another plan would initiate the creation of a waste-management system. In this plan, low educated, unemployed Romani (and non-Romani) could find employment by selecting waste under controlled working conditions. The products could be sold to aggregates, which would then reutilise diverse waste such as pet bottles, rubber, etc. in order to generate energy. Additionally, a reparation workshop could contribute to the reutilisation of repairable tools found in the waste.
3. Finally, under the leadership of local Romani leaders, marginalised Romani village communities could be upgraded and hygienic standards increased and maintained, similar to the clean and established villages and small towns of the region. This would be achieved by the self-organising of Romani communities.

The sub-projects for Renewable energy systems (RES) technology-based municipal energy plants were the first to be realised. On the contrary, the realisation of the Romafa was still waiting for resource-strong stakeholders, at the time of this research (February 2012 to May 2013). No municipalities or private entrepreneurs seemed ready to support the Romani Minority Self-government (RMS or in Hungarian CKÖ) initiatives. Instead, the municipalities asked were satisfied with the current arrangement for waste management. Three biogas aggregators were in the phase of 'projectification'. Two of these planned units were to be arranged in collaboration with several municipalities and they planned to introduce a coupon system, though none of the planned plants were envisioned to be connected to the electricity supply of marginalised housing areas and were not to be placed adjacent to Romani settlements.

By focusing on the Romafa project's efforts to incorporate social aspects into the utilisation of renewable energy, this chapter will contribute to our understanding of how different interests influence the targeting and realisation of developmental goals, aimed at improving the living conditions of marginalised groups, and whether and under which conditions the new model of governance can work, on the local level, for the benefit of those with the least resources.

Theoretical Background: Partners in Rural Development?

In contrast to traditional models of rural development, in which governments promoted their preferred developmental agenda through market mechanisms, LEADER is considered to be a potential instrument to facilitate the re-embedding of local societies. According to Kearney et al. (1994, 22), the LEADER concept refers to the capacity building of the entire local population aiming 'to create public goods that will help to overcome the instances of market failure which characterize rural economies'. In the LEADER process, stakeholders representing multiple sections of society, such as firms, municipalities and civil society, compete for the available resources of the quasi-market.

The question for this chapter is: to what degree are the concerns of social sustainability, one aspect of which is the situation of marginalised groups, commensurate with other socio-economic interests in the realisation of Romafa's goals? This issue is problematised through two aspects.

Firstly, the key stakeholders (firms, municipalities, civil society and households) in the LAG are identified and the question posed: to which degree did these different segments of society work toward the goal of improving the living conditions of the target group? Polanyi's classical categorisation of the various types of economic coordination principles is used here to help understand divergences in the interest of motivating local actors. Simplified, the four actor types are the firm, the local state, civic society and the household, because they differ according to their basic interests (Asztalos Morell 1999; based on Polányi 1976). Söderbaum (2011) continues this analysis by differentiating between concurrence-oriented- and collaborative models. The first type is based on self-interest, with a focus on profit, and the survival of the unit assumes an extended reproduction. The latter type is based on the principle of care for others, and its aim is to achieve benefits for all the members within the network. Moreover, its survival is measured in how it can maintain mutually beneficial contacts with its members (Lázár 2010; Frank 2004). Whereas the welfare-states work along the constituency of a broad citizenship, civil society actions are formed around partial interests. The proponents of sustainable development argue that a socially and ecologically sustainable society assumes collaborative models and ethical concerns are to be incorporated not only among the strategies of idealistic organisations, but even with profit-oriented companies (Söderbaum 2011, 49–50).

Nonetheless, even some idealistic organisations can be driven partly by commercial goals although commercial and collaborative interests are not by necessity reconcilable with each other. In a rural context, alternative initiatives that stretch beyond what the market and the welfare state offer emerge, striving for the establishment of autonomous and self-sufficient units for those in need, such as systems for food sovereignty and community-supported agriculture, based on the decommodification of food and land (Dyck 1994; Lázár 2010).

Secondly, the model of participatory redistribution assumes that those engaged with developmental processes have both social and cultural capabilities (Sen

1984) to manipulate and work for the definition of development targets and their realisation (Shucksmith 2000; Shortall and Shucksmith 1998). Different access to resources, with Bourdieu's (1986) terms of material and immaterial capital, is the source of social inequalities. Since the resources and capabilities to influence are not equally distributed within the communities, the ability of marginalised groups cannot be taken for granted. Marsden (2008) argues that the strength of networks within communities is intertwined with the potential of communities to grow. Thus, we could argue that communities split along social, symbolic and cultural ruptures form a hindrance for the improvement of the conditions of those on the periphery. Two aspects of the improvement of the conditions of marginalised groups is utilised in this analysis. On one hand empowerment is seen, following Narayan (2002), as the 'expansion of assets and capabilities of people to participate in, negotiate with, influence, control, and hold accountable institutions that affect their lives'. On the other hand, the concept of capacity building is used, in which the focus will be on the accumulation of social capital. Social capital building is perceived following Putnam (2000) to be associated not only with the individuals' access to powerful networks (Bourdieu 1986), but also with the relations of trust within a community that allows for the facilitation of cooperation for mutual goals.

Thus, the second question posed in this chapter is; in what way has the participatory model assisted in realising a bottom-up process, democratic participation and the empowerment of resource-poor participants in the project?

Regional Enclaves of Impoverished 'Rust-pockets'[2] and the Ethnification of Poverty

The Miskolc region, where the FA LEADER is located, is one of Hungary's crises areas. The per capita income level in this region is 50 per cent of the national average, while the proportion of new investments lies at 20 per cent of the national average (Cserehäti 2010). The Miskolc region had been one of the heavy industrial and mining centres of Hungary during state socialism, employing large sections of the rural population with a low level of education, a large proportion of whom were of Romani origin and many of them were commuting, low-skilled black-collar workers (Kemény 1976, 2004; Ladányi and Szelényi 2003). The transition to capitalism led to the mass closing down of the industries and mines as well as the dissolution of agricultural cooperatives (Kovách 2010). As a consequence, former industrial areas were transformed to socially marginalised crises areas.

2 The term 'rust-pocket' was introduced by Ladányi and Szelényi (2002). They refer with this term to such marginal small communities and regions, where the postsocialist transition led to the shut-down of low skill work opportunities, which were found in mines, heavy industry, construction and agriculture, resulting in enclaves, i.e. rust-pockets, with high unemployment and social exclusion characterising a long-term underclass.

Unemployment hit the Romani residing in the region hardest. Unemployment is most concentrated in so-called 'rust-pocket' regions of the country, where small village societies are the most affected (Kovács 2008). Lacking skills, they could not reintegrate into new branches, which required more diversified skills (Emigh, Fodor and Szelényi 2001, 3). This marginal situation is passed down to subsequent generations, since the school system is not capable of lifting children out of poverty (Molnár and Dupcsik 2008). According to a survey, which included 1,500 households in the FA LEADER region, 83.9 per cent of Romani in the age group 19–34 were inactive compared to 49 per cent in the Hungarian population at large (Csereh"ati 2010, 52). The main source of incomes for these families is social benefits. Inactivity is associated with low incomes and poverty.

The majority of Romani (2/3) lives in mid-sized rural towns or in small municipalities, of these, 13.1 per cent live in sub-standard 'need housing' (*putri*). One-fifth of the Romani reside in ethnic enclaves which are sizeable, even if this degree of concentration is low compared to neighbouring countries (Ladányi and Szelényi 2004; UNDP 2005).

In a process that Váradi et al. (2010) identify as counter-selective mobility, resourceful members (both Hungarians and Romani) leave aging communities due to lack of employment opportunities. Meanwhile, these communities become the targets of mobility for social strata weak in resources. This counter-selective mobility has obtained ethnic dimensions. In these communities previously established harmonic forms of ethnic co-existence (Szuhay, 2005) become challenged, and often the original Romani residents associate themselves with the local Hungarians in opposition to the newly immigrating Romani (Durst 2008; Schwartz, 2012; Kotics 2012; Szabó-Tóth 2012). On the one hand, impoverished communities recreate dysfunctional cultures of poverty (Ladányi and Szelényi, 2004; Durst, 2008), whereas on the other, the discriminatory practices of the majority society contribute to the reproduction of marginal positions (Schwartz, 2012).

Ethnic Diversity as a Challenge for Rural Governance in Hungary

The so-called municipalisation of welfare (Szalai 2007) has been identified as one key structural explanatory factor for the reproduction of poverty. The re-instalment of the autonomy of small municipalities resulted in the creation of resource poor units burdened with social responsibilities without adequate resources. The Hungarian Government launched a new regional development policy in 1996 signifying a turn towards the EU trend. With Hungary's joining the EU, the power relations of local resource distribution were rearranged and became multi-participatory (Kovách 2010, 174–177). Projects available through a quasi-market distribution became the primary means of accessing resources, leaving behind the former system of support that guaranteed revenues calculable on universal principles. The projectification of rural development can be considered as an accompanying symptom of the neoliberal turn in welfare systems, which was signified by the retrenchment of the universal benefits of Soviet-type welfare regimes toward means tested systems (Kay 2010; Haney 2002) combined with intensified moralisation of the importance of work. Just like in the EU, the new style of governance is characterised by a plurality of interests, actors and networks, strengthening the role of local particularities in local welfare regimes: 'poverty, as social problem becomes a small community issue ... the conflict between poor and not poor appears as the malfunction of the local communities' (Szalai 2002, 39). Furthermore, the decentralisation of welfare to the community level gave power to local officials to negotiate entitlements to benefits. These negotiations lead (Thelen 2012) to differentiation among the deserving and undeserving poor, which often become ethnically contextualised (Schwartz, 2012; Hamar 2010) and stand as grounds of entitlements.

The importance of civil society increased in the functioning of local welfare regimes in the new models of governance. Soviet-type systems aggravated the condition for bottom-up, empowering civil organising (Rose 1998), the importance of which for local development has been lifted up in diverse contexts (Krishna 2010; Svendsen 2004). Sätre (2012) found in her study in Russia that these legacies fill a function even in today's society, since NGOs assisting the poor are most often formed top-down rather than bottom-up, and rather than working for empowerment stand for helping with doles.

In Hungary the institution of Minority Self-governments (Molnár 2003; Molnár and Schafft 2003a and b) secures a system of local and national representation of minority interests even if they lack sizeable resources. Nonetheless, Csongor and Lukács (2003) indicate that while the overwhelming majority of financing for pro-Romani projects originates from diverse state sources, civil organisations, as well as the Romani Minority Self-governments (RMS), play an active part in facilitating project applications that allow the utilisation of these funds. The capabilities of Romani civil organisations is increasing not least thanks to the presence of international foundations contributing to the dissemination of Western-type, local empowerment-oriented development methods. (Csongor and Lukács, 2003, 49) However, many of the most marginalised communities typically lack elites on the local level (Vajda 2008) or if there are such elites they may distinguish themselves from the poor, and projects aiming to empower communities do not reach those in need (Molnár and Dupcsik 2008).

Renewable Energy and the Ethnification of Energy Poverty?

Energy-poverty in developing countries (Bierbaum 2011; Casillas 2010; Cecelski 2000) is associated with the lack of access to quality energy services. In contrast, in Central and Eastern Europe, one talks instead about fuel poverty, referring to the inability to afford energy, rather than the non-existence of services (Herrero 2011). Despite the fact that the energy consumption per person in Hungary is among the lowest within the EU, the proportion of energy expenses within household expenses are relatively high (Kajati 2000; Hegedüs 2007, 74). Firstly, households have low energy-efficiency (buildings and equipment) (Buzar 2007). Secondly, the energy supply system has been privatised to global companies: in 2005 91 per cent of electricity was supplied by three international monopolies (E-ON, RWE and EdF), which achieved regional monopoly positions (Kajati 2011, 108; Járosi 2007). Privatisation, and gradual price deregulation (Kaderják 2010, 8) was accompanied by a four times higher increase in energy prices than average consumer prices (KSH 2010). Thirdly, the urban-rural poverty gap has widened since the 1980s (Szoboszlai 2004), while small regions with a high concentration of unemployed inhabitants occupy an especially disadvantaged position (Váradi 2008), which even impacted on fuel poverty. Renewable energy systems create a technology that can also be an independent energy producer, and through this contribute to energy sovereignty. However, as soon as the energy leaves the producer's unit and is ready for transport, it turns into an exchange relation with the major supplier system. Therefore, efforts to increase the energy sovereignty of marginalised Romani communities has to be analysed in the context of local, regional, national and international relations into which energy systems are embedded.

Research Design

This chapter focuses on two sub-projects within Romafa: biogas aggregates and the waste-management utilisation system. The 44 municipalities form an important stakeholder group in FA LEADER. None of these communities showed an interest in the realisation of the waste-management project, while three have worked out detailed project plans and filed applications to FA LEADER fund competitions with the intention of installing biogas units. In two of the municipalities the politically non-aligned mayors initiated the project, whereas in the third a *Jobbik* (right-wing radical) party politician was responsible for the biogas project. A fourth municipality, with an MSZP (Socialist Party) mayor is also of interest, since the local Romani RSG leader together with the regional Romani *Vajda*[3] were the initiators of Romafa; even so, the project has not been realised in their municipality. Three of the municipalities, B, C, D, have over 7,000 inhabitants and count as small towns, while the fourth municipality, A, has approximately 2,000 inhabitants. The proportion of Romani is between 20–30 per cent in these settlements, while 80–90 per cent of the unemployed are of Romani origin. Among the Romani residents of these four municipalities there were also better-off Romani families, who live either integrated into the main village, as in municipality A and B, or are living within the segregated poor Roma enclaves, as in municipality D. However, the target group for Romafa were the inhabitants of poor marginalised enclaves.

23 interviews with 15 stakeholders were conducted, representing different societal organisations and interests (firms, municipalities, civil organisations, residents) in four municipalities during four field studies in rural communities in

3 Traditional Romani leader.

the Miskolc region between April 2012 and February 2013. The interviews were from one to eight hours long. Among the interviewed, there were municipal leaders from four municipalities, the director of FA LEADER, one politician involved with one of the biogas units, three ethnic Romani and one ethnic Hungarian inhabitants in a marginalised settlement within a rural community characterised by high unemployment, one of the RMS leaders of a larger Romani community, a traditional Romani leader, a *Vajda*, and lastly one ethnic Hungarian and five ethnic Romani entrepreneurs. Four of the key informants were re-interviewed two to four times each in order to validate the information. Below, I present the development from the perspective of those who belonged to the sphere of interest of the FA LEADER consortium, which implies that the voices of stakeholders outside the consortium such as large waste-management companies and electric companies are not represented.

Key Stakeholders in the Civil Society Initiated Waste-Management Project

There are two Romani civil society leaders representing two very different Romani organisations active as stakeholders in FA LEADER. Attila Kardos is a County *Vajda*, a traditional Romani chief, in municipality B. According to Romafa (Interview 1), under the leadership of Kardos, a non-profit company was to stand behind the waste-management project. The other representative promoting Romafa both locally and regionally is László Rózsa, Romani Minority Self-government leader in municipality D. Rózsa proposed that waste would be collected by carriage and pair, providing frequent service and work opportunities, while saving the roads from the damage caused by large trucks collecting the waste as they did at the moment (Interview 2). This project presupposed collaboration with municipalities due to expected financial support, and since community waste management is a responsibility delegated to municipalities by the state, municipality D was approached to become the centre of a waste-sorting plant, thus creating additional work opportunities.

However, Rózsa and Kardos have not been able to mobilise support for their RES projects, which was motivated by different arguments. Firstly, the municipalities of the region formed a consortium and entered into a long-term contract with a large-scale company for the collection of the garbage. This company was to run disposal sites according to EU standards, while municipalities would reimburse the costs from the local inhabitants in the form of taxes. The mayors interviewed were satisfied with the service. Introducing the waste-management ideas of Romafa would have challenged this collaborative system.

Furthermore, the municipalities themselves had limited resources to mobilise and chose to prioritise other investments. 'I do not know which community's interests the mayor represents?', argued Rózsa (Interview 2), feeling that the mayors take the side of the interests of large companies, rather than promoting the interests of low-educated social groups with their limited resources and opportunities for the generation of values for themselves by the utilisation of resources, such as garbage. Meanwhile, the resource poverty of Romani organisations prevents them from being able to access loans that could enable them to complement their own EU funds, which require prefinancing. LEADER was criticised for favouring big funds, which is contradictory to small local projects. Funding should be more flexible to allow poor communities to participate without a rich sponsor.

Finally, the mayor of municipality A (Interview 3) argued that the working conditions for those sorting the garbage would not provide respectable employment opportunities. Hence, this mayor challenged the vision of fitting the poorest technology to those with the least resources. She also raised concerns as to whether the idea was to utilise the vulnerabilities of the people employed. In contrast to the mayor, the Romani Minority Self-government leader Rózsa (Interview 2) framed the planned activity in the context of the emerging RES technology: 'It is not a question of humiliating work; instead, this is truly alternative energy production that we were to base on waste-management economy'. Thus, in his mind, combining waste management with RES technology would raise the status of the work. This dissonance brings up the issue of who should define the acceptable conditions of work: the Romani leaders, the municipal leaders or the prospective workers?

These difficulties indicate weaknesses of the LEADER concept for pursuing the interests of marginalised groups. The ability of civil society agents might be curtailed by the lack of support from municipalities by not providing necessary support and resources.

Romani Representatives as Stakeholders of Civil Initiatives

Especially difficult is the situation of resource-poor, stigmatised minorities to capture initiatives. As representatives of their communities, Kardos and Rózsa have to balance supporting Romani of a given settlement with act within the larger society. Navigating between these two spheres, they utilise different discourses of marginality. Being Romani is central to their identity in a social and cultural sense. In the same time, they join with some elements of the social conservative discourse, taking a critical position towards the Romani. When interviewed, Rózsa is critical of what he sees as a downward development in Romani communities. He identifies garbage and 'survival criminality' as key issues contributing to stigmatisation of Romani, both having their roots in poverty. Since households cannot afford waste bins, waste is thrown away in the gardens or around the village. Kardos and Rózsa see the roots of apathy in long-term unemployment and existential insecurity. They see criminality and apathy as characteristic of one segment of Romani communities: 'To be honest, those poorer, go away to steal. This is not debated and it causes problems' (Interview 2).

Rózsa (Interview 2) argues that this downward spiral can be broken if efforts are made both within Romani communities and by the majority society. Working in the community, where poverty is a shared experience among Romani and non-Romani alike, Rózsa articulated poverty as the major entitlement grounds rather than ethnicity, and argued for 'not putting forward the interests of only one folk group [i.e. the Romani], but of everybody living in the periphery'. Although Rózsa did not succeed with his ideas in Romafa, he gained recognition for a Social Foundation financed waste project, which employed 25 long-term unemployed participants last year, out of which 15 were Romani and 10 were local Hungarians:

> I think that we live together, we go to the same hospital, and we get on the same bus. It is my principle that we should not divide it, since this would increase hatred. That now it was again the gypsies who received something. … I am just as poor as you, despite being Hungarian. And we do not receive because we are not gypsies. (Interview 2)

The new forms of participatory governance operate through projects. To get a project, applicants have to possess substantial cultural (habitus), human (skills) and social capital. There is, however, a capital deficit among those without resources, compared to the others possessing immaterial capital assets and being able to transfer them to material resources through projects, and possibly to obtain a stronger position within the redistributive system. Lacking the required material resources as well as local intelligentsia capable of initiating projects, the ability of groups without resources depends on the willingness of sponsors and funds to assist them. Neither the Romani Minority Self-governments nor the *Vajda* have earmarked means to finance applications. The *Vajda* says he does not to derive any income from his position, to the contrary, he claims to show his engagement for the community by standing up for those in need. The limitations inherent in the institution of the *Vajda* are described by Láng (Interview 4) in the following way: 'He has a good will to act for the benefit of the community. But he has neither horses, nor money, nor weapons'. Even if he had private means, he does not have the support of the state, since the state acknowledges the legitimacy of the democratically elected Romani Minority Self-governments (CKÖ) representatives, but not of the *Vajda*. Even if CKÖs enjoy a higher institutional legitimacy, the CKÖ representatives do not manage sizeable resources either. As Rózsa (Interview 2) explains, when they founded a social cooperative, they collected funding from 10 private Romani persons in the community who could contribute 10,000 Forint (€34) each. They had to finance the expense of the lawyers arranging the proper registration forms, so when they applied they registered the expenses of the administration as the cooperative's own input. Although the board has voted a fee for him as a leader of the cooperative, he has not taken any of the money, so to better channel all the resources for the project's applications and realisation.

Economically strong organisations, such as firms or (most) municipalities, can afford to buy the services of expert project firms, or to have such experts on their payroll. The preparation of project applications assumes not only know-how about the application process, but also know-how about the specific field of application. The realisation of the Romafa project assumes, beyond technological know-how on renewable energy, entrepreneurial know-how and a socially innovative approach to its utilisation for community development. Since there is no established model for such an operation, the importance of experts is even more central.

The Waste-Management Project and Capital Interests

Romani Minority Self-governments' leaders attributed some of the difficulties experienced while attempting to realise the project to the concurrent capital interests. On the one hand, different stakeholders (e.g. Láng, Interview 4) within the project indicated that the expansion of renewable energy products is to be realised in an electricity market, in which a few companies have monopoly-like positions. Electricity companies are not interested in contributing to the development of the self-subsistence of municipalities, communities or households. On the other hand, there are capital interests within waste management, and capital-strong agents enjoy economic advantages from their contracts with the municipalities. Their interest is to maintain these contracts, and consequently, not let waste be utilised for the enrichment of local communities.

In challenging the established system, one needs both economic capital and membership in influential social networks. The utilisation of diverse valuable side-products assumes a specific technology, which further assumes larger capital investments. To start with, FA LEADER was to realise its plans in collaboration with the B *Vajda*'s non-profit organisation. In 2012, they came in contact with the successful and capital strong Romani entrepreneur Albert Balogh (Interview 5). He himself started his career in an iron recycling station as a blue-collar worker, climbing the ranks through hard work. He leads his operation according to a profit maximising principle, and prior to his contact with FA LEADER he invested in a factory that could generate different final products such as energy, briquettes, warmth and nano-technological production through advanced pirolization technology. This production has a large profit-generating potential. In order to put his factory into operation, Balogh needs collaborative partners, and FA LEADER can assist him with this. One result of this contact is a new collaboration with university researchers who can help him with the finalisation of the technology, with the profit potential in this field helping to catalyse interest in the realisation of the project. However, LEADER cannot provide the economic support needed.

Investments in profit-oriented establishments are motivated by expected benefits and the minimisation of risk. The question is how capital interests can be combined with both social goals, such as that formulated by Romafa. Is it reasonable to expect a Romani entrepreneur to exhibit a higher consideration for impoverished Romanis' opportunities to generate energy from waste only because he is also of Romani origin? Perhaps this scepticism finds expression in one of the interviewee's views: 'As soon as an entrepreneur becomes successful, he turns against his community. His job is not to drive a half-financial operation, but to maximise his profits and make use of laws, regulations and opportunities, a community's lack of knowledge and ungainliness'. As the interviewee continues, this just could be the reason why one might distance himself from the impoverished if one has succeeded in breaking out of such a community: 'He will no longer work for the betterment of the community. He would cut ties with his origins. The interest of the individual and of the community departs'. Thus, one should not take for granted that successful Romani entrepreneurs will manifest a larger commitment towards the interests of marginalised Romani communities compared to non-Romani. This corresponds with what both Bourdieu and Swain argue in relation to the nature of social contacts with kin in marginal communities: 'Family relationships of friendship, support, a helping hand in time of need, are not social capital. They are 'shackles' to be broken' (Bourdieu 1986, 389, see this emphasised in Swain 2003, 189). The experiences from the Romafa project challenge this perception. Strong ties within marginalised communities do provide ties of solidarity. While Romani workers experience discrimination on the labour market, Balogh is providing working opportunities for them, since 80 per cent of his employees are Romani. In opposition to the dominant view, his perception is that Romani are very highly motivated to work and are ready for higher inputs compared to non-Romani and are also ready to work with high-risk tasks. This latter aspect, however, indicates also the potential vulnerabilities of those employed.

Moreover, profit-maximising operations can be combined with the social goals of Romafa. Balogh's plan is to arrange waste collection through the placement of selective garbage bag containers in which households could selectively place their waste into bags collected by trucks. In opposition to the conditions of today, he would collect the selected garbage, and could even offer some refund. Hence, those who today have to pay for getting rid of their garbage could make some income from it. Nonetheless, he expects to be able to make an extra profit if his factory comes into operation since then those who collect and utilise the selected garbage would be the same, with no middleman in between. Collecting the trash bags from the individual households with large and heavy trucks assumes many stops, which would increase his energy expenses. This could motivate him to collaborate with the Romafa project without compromising his profit-maximising interest.

Key Stakeholders in the Realisation of the Biogas Project

As of 2013, there are three biogas generators which have accepted detailed project plans and have filed applications to FA LEADER tenders. Two of the projects were initiated by municipalities, and one by a *Jobbik* (right-wing radical party) politician (Interview 6).

Municipalities A and B are to build 60 kilowatt biogas units, each with a value of 130 million Forints (€44,000), in collaboration with neighbouring municipalities. According to the mayor of municipality B (Interview 3 and 7), the unit is planned to work as a non-profit company of the municipality, which is estimated to give employment to six persons each. Both of the planned units aimed to utilise the coupon system and waste delivered by local citizens would either be paid in cash or by coupons which could then be refunded by the energy supplier in the form of electricity. The mayors of municipality A and B are expecting a positive outcome of this system. They would no longer have to use public work for keeping the village tidy, the energy or the value from it would cut municipal expenses: 'For me it would be pure happiness, even if it would just help to pay for the expenses of public lighting'. Meanwhile the unit is expected to create enough value that it can 'finance the employment of people by what the plant is going to produce'. Additional incomes will benefit the two municipalities. Beyond producing electricity, municipality B plans to utilise the heat produced in the process for greenhouse production which would supply vegetables to the municipality's institutions.

The third planned biogas plant is to be located in municipality C. In opposition to the two others, this project was not initiated by the municipality leaders, but by a *Jobbik* party politician (Interview 6). He argued that the development of the coupon model would require an advanced system that would reach beyond what he could administer. In the form of a new non-profit company allied with the municipality, the plant will utilise all of the green waste generated in the municipality's public areas which would be collected by public workers. Public work relation has the potential to create win-win collaboration between the municipality and the non-profit producer; meanwhile it is a disadvantageous work form for the employed compared to a 'real' work contract. The non-profit company running the plant was envisaged to be open for market-oriented production in order to maximise the production potential of the generator. The income of the operation would be decided on by the company board and is planned to be donated to local NGOs and would be directly utilised by municipal buildings on the site.

Thus all three project proposals prioritised renewable energy technology with economic benefits for the municipalities and in the last case for NGOs and only marginally for peripheral communities. There were no strong capital interests bound to these projects within the LEADER group, which might have contributed to the fact that none of these three stakeholders succeeded in subsequent LEADER tenders to acquire funding.

LEADER Management and Local Initiatives

The main entrepreneurial and innovative mind behind the FA LEADER concept is Antal Láng (Interview 4). Without his efforts, the LEADER group would without doubt never have been created. He, along with the management team surrounding him, has made a large impact in introducing RES technology at the local, regional and even national level in Hungary. The realisation of these business ideas presupposes widespread, influential networks and collaboration in order to lobby to convince regional planners and to compete with global actors. He has been the leading person in creating the regional network of the FA LEADER, which includes municipality leaders, entrepreneurs and NGOs. To compete successfully for development sources assumes a professional team occupied with project management. The expert project team recruited by Láng continuously maps the interests of local stakeholders, while also initiating and informing them about new development ideas. Thus, Láng has had a central role in regional capacity building for the projectification and implementation of RES-based technologies.

However, beyond being the visionary for RES technology for rural development, Láng is also a successful entrepreneur in his own right and has been a driving force in spreading this technology and his own entrepreneurial interests. One of his private visions is to introduce an RES energy-based transport system into the region, which requires the local production of energy, a system for distributing it to consumers and the availability of vehicles to utilise it; thus, he has own financial interests in the competition for LEADER resources.

During the first phase of the existence of FA LEADER, municipalities succeeded in realising several RES technological investments financed by LEADER, such as solar power collection and wind power stations. However, the continued realisation of municipal- and NGO-based projects stalled. The waste-management project could not even reach the stage of proliferation to prioritised projects due to collision with capital interests outside the LAG and the lack of support by municipalities allied with these interests. Láng's strategy was to lobby for the project, together with the Romani Minority Self-governments' leaders at regional forums. However, after being hindered, Láng has turned to establishing collaboration with capital strong investors in the waste branch-and abandoning the Romafa concept.

In contrast, the biogas unit projects could not be fed up to the application platforms, despite having been accepted by the consortium in the process of proliferation as valid priority projects. Due to the scarcity of resources, the consortium decided to close down the electronic applications sides when the value of the applied projects reached the level of available financial resources. This created an advantage for those 'close to the fire', insofar as knowing when the announcements were to be made. Some consortium members said disappointedly: 'We are only a frippery to FA LEADER. They are too big, and we are too small to be able to realise our interests'.

LEADER LAG Strategies toward Peripheral Groups

Even if the Romafa project envisages the creation of a self-managing energy supply, it is not planned to force it on the communities; instead, it is expected to work under a self-organising principle that people are free to join, its planning and realisation efforts are not those of participatory democracy. Rather, the manager of the consortium, Láng proposes a clear top-down developmental model driven by an expert team referring to the project as 'a flag under which people can be organized into rows'. Even if Romani leaders were involved in its development, the concept is not the result of a bottom-up process, thereby facilitating the direct participation of the marginalised groups that the project is targeting. Láng argues that the preconditions for the project presuppose strategies and economic, political, social and cultural assets that are simply inaccessible for the marginalised enclaves.

The attitudes toward the realisation of social elements in renewable energy sources -projects in the municipalities studied could be categorised into different models drawn along the degree to which socially conservative and socially sensitive discourses mix concerning their practice towards socially marginal groups.

As discussed earlier, Rózsa, the local leader of Romani Minority Self-government (RMS), who was the most active in the initiation of the Romafa project, in addition to the BAZ *Vajda* Kardos, comes from municipality D. Here, there is an on-going overt conflict between the Romani community and the municipality. The municipality leadership has prioritised the mobilisation of resources for the renovation of all municipal buildings, while they have not initiated any rehabilitation or social development project in the two marginalised Romani enclaves. The mayor (Interview 8) was satisfied by stating: 'It is difficult to make people change their attitudes', a typical social conservative attitude blaming the poor for their poverty. Rather, it was the local RMS leader Rózsa, who has been successful in applying for EU and national funds for a waste reclamation project and a social cooperative project mobilising long-term unemployed Romani residents.

In contrast to municipality D, the socially sensitive approach dominates in municipality A, the mayor of which has initiated diverse programmes for marginalised groups. One of her instruments is public work. However, the municipality did not qualify for preferential level state support for the unemployed due to belonging administratively to the Miskolc district, which on average has a lower unemployment rate than is stipulated to be eligible for support. Out of 2,000 inhabitants in municipality A, 190 are unemployed, and out of these the municipality can provide public work for only 20. Her aim is to create value-generating occupations. One of the work teams produces ecological food, and another team produces pasta to be supplied to the local school. Production is combined with training for the workers in ecological production. Such programmes should employ both Romani and non-Romani alike. According to the mayor, such activities are not viable in a market, and turned out to have a zero deficit since the work expenses are covered by diverse state support. However, she feels that these jobs generate social profits, as they give meaning for the participants and produce higher quality goods for the school children. Finally, those participating learn new knowledge particularly among the marginalised Romani participants. Her activity could be described with the term social innovations (Nikula et al. 2011).

My interviews with local municipal mayors also shed light on how the municipalities worked under the conditions of a lack of resources. The implementation of projects within the framework of the LEADER project assumes prefinancing, which means that the municipalities must prioritise between their different projects. The prioritised investments have often led to projects which have contributed to the decrease of the municipality's own expenses. Hence, the action or lack of action on the part of the municipalities is formed in the broader context of economic constraints and the understandings of those in power on the causes of the vulnerabilities of socially marginalised groups.

The lack of resources also strengthens the practice of categorising the poor into deserving and undeserving, a praxis that was most clearly articulate in case of municipality B. Here out of 400 unemployed people who should be eligible for anti-unemployment interventions, state resources allow only the employment of 55 persons through diverse measures. Most of these unemployed are Romani (90 per cent) with low employability as a consequence of a low level of education and skills. Meanwhile, the mayor (Interview 7) differentiates between the marginalised who are striving to be integrated, referring to the Romani members of a local Methodist congregation and those who are perceived as ‘only waiting to be taken care of by society’. The members of the Romani Methodist community argue that there are only a few trouble-making Romani families who contribute to atrocities, such as the recent beating up of an old Hungarian man in his garden by Romani youth. The congregation organised a protest march against such atrocities forwarding an apology in the name of Romani to the inhabitants of the municipality (Kovács and Zubor 2013; BGY 2013). Meanwhile, the community was the site of a shooting (Tódor 2012) of a Romani young man fired by ethnic hatred. Thus the association of ‘being undeserving’ is practiced with the Romani as a group, yet it is at the same time highly contested.

Economic constraints set limits on projectification for the benefit of vulnerable groups even when the political will exist. Those, who could finally benefit from the various projects realised by the municipality initiatives were not involved in the project planning process. Thus, they cannot be considered as ‘empowered participants’, in the meaning defined earlier by Narayan (2002). Instead, they are chosen as ‘deserving to be poor’ by the authorities for the tasks. The mayor argued that project applications all too frequently were not resulting in funding, and that those in poverty would just be confused and she would lose her credibility if she planted expectations, the realisation of which were beyond her control. Thus, even if she, in opposition to the mayor in municipality D, had a social sensibility towards the conditions of the marginalised, she acted as a benevolent ruler in a top-down fashion rather than as a democratic ruler building bottom-up processes.

As for capacity building, the cultural capital transferred by municipality projects is highly contextual. The acquired knowhow does not increase the employability of participants on the labour market, while the disciplinary and punitive aspects of the work organisation of public labour discourage autonomous agency. In contrast, the proposed model for the biogas unit could be considered innovative since, in opposition to the model offered by public labour, its purpose was to create ‘real’ wage work. Furthermore, by the utilisation of the coupon system, it was to rely on the self-interests of those delivering raw material to the plant and by this it was to motivate for autonomous agency. Therefore, even if on a small scale, the project was intended to contribute to capacity building.

The relation between civil society organisations and marginalised groups is similar. Those most vulnerable groups who succeed in becoming parts of projects benefit in some form or another from these. Even so, they do not become part of the project planning phase. Therefore, their capacities to participate in development projects do not progress, and no empowerment process is launched in this respect. Capacity building in community development remains restricted to the quite few activists driving the projectification process. To incorporate those with the least resources, i.e. the marginalised Romani communities, assumes an increased level of civil society participation in the communities (Kotics 2012). Efforts to alleviate poverty should also focus on the social capacity building of disadvantaged groups, which assumes that competencies accumulated in social research (Kotics 2012) as well as in established civil society organisations, such as Autonomia (Kovács 2012), are given a platform to operate from. The state at large and local municipalities in particular, would need to take on the responsibility to promote the strengthening of material and immaterial resources that can contribute to the improvement of the capabilities (Sen 1984) of the marginalised groups to be able to help themselves out of poverty. Decentralised models of participatory governance seem to have deficiencies in meeting these demands.

Conclusions: Failures and Potentials of the Romafa Project

The experience of Hungarian FA LEADER thus supports the earlier research outcomes in EU context concerning the shortcomings of the LEADER concept in empowering the least resourceful groups (Shucksmith 2000; Curtin et al. 1997; Macken-Walsh et al. 2012). Although LEADER provides new ways of cooperation and assists in the accumulation of social and cultural capital in the different segments of the LAG, LEADER does not alter in fundamental ways the underlying principles motivating the different types of participants in the LAGs. As the example of FA LEADER indicates, the stakeholders rich in resources, out of different motivations, disposed of most of the resources required for the implementation of technological advances. The motivation of LAG members differ according to the type of economic organisation they represent, which fits well with Polányi's theory of forms of economic coordination. Functioning within the paradigm of the market, firms are motivated by an interest in profit maximisation and stand in competition for the control over different segments of the market and resources. An alliance with the interests of the poorest is not by necessity incommensurate with an ability to realise profit, but is by far not adrift by itself. Without such an alliance, those who are the poorest in resources do not have the necessary economic, social, cultural and symbolic capital to enable them to make claims on local opportunities to enter the market. Although municipalities are responsible for social welfare issues, they are also curtailed by a lack of resources, in which they have to prioritise between different developmental targets, where keeping their own economy in balance becomes a goal in and of itself. Furthermore, many of the municipalities within the FA LEADER are split socially and culturally with marginalised enclaves. As Marsden (2008) argues, socially split communities do not provide favourable ground for development. In FA LEADER, as in many other previously studied LEADER communities, local notables utilise the opportunities, while the exclusion of symbolically stigmatised groups is reinforced.

In Hungary, Romani Minority Self-governments have provided institutional frameworks for the representation of local Romani interests. However, Romani leaders balance between identifying with the interests of those with the least amount of resources, and building their own bridging social capital (Putnam 2000) through relations with the local and national economic and administrative elites. Having roots in marginal communities enables them to mobilise this 'barefoot' army of reserve labour when opportunities emerge. Nonetheless, they seem to operate most efficiently in contexts where they can utilise the synergies between their roles as entrepreneurs and representatives of the marginal group. They work on extending their capacities through social and cultural capital building within the LEADER group. Even so, it appears that the increase in these capacities can only partially compensate for the deficit in symbolic capital characterising the group they represent in the local socio-economic context. They seem to be able to manoeuvre more successfully through their national level contacts, and by bringing home alternatives that do not presuppose collaboration with local authorities. Even if the Romani elite has not realised the goals of Romafa, it needs to be noted that by promoting the Romani elite, first of all by Romani entrepreneurs in the region, FA LEADER indirectly contributes to the advancement of work opportunities of Romani in the region. As Láng (Interview 4) indicated, when an entrepreneurial group receives resources for development, also the broader community benefits and finds new opportunities.

LEADER has evolved as a model of participatory democracy. Nevertheless, the deliberation of projects, participation in calls and the realisation of the projects assume extensive cultural, human and social capital, as well as resources, over time. Those possessing economic capital can buy the services they need, while the poor have to rely on their representatives, who themselves often do not have access to the economic means necessary to finance expert services. The LEADER centrum fulfilled a capacity-building function towards those actors who to some degree already had economic, social and cultural capital. Although they contributed to the formation of the Romafa concept as part of their strategy document, they have not been engaged with direct capacity building towards the peripheral Romani communities. Instead, they made the assumption of steering Romafa in a coalition with Romani leaders and municipalities, i.e. into a top-down model.

It also seems that the ideas of FA LEADER for improving the conditions of the marginalised group assume that those with the least resources lack the ability to meaningfully participate in the deliberation of ideas concerning their conditions. Rather, ideas for their benefits are constructed in collaborations between the leadership of the LAG together with the LAG member RMS leadership. These act as benevolent rulers who work for the benefit of those without resources in a top-down fashion, which does not contribute to the building of the capacities of the poorest to articulate and proceed autonomously in realising their own goals. The involvement of marginalised communities may have enabled the local Romani leaders to put pressure on municipal leaders to support their initiatives. Thus efforts to alleviate poverty need also focus on social capacity building. This effort should engage those beyond NGOs, even state and municipal actors and institutions.

Finally, the precondition for the realisation of LEADER projects is that the participating LAGs are able to finance their project activities and will be reimbursed by EU only afterwards. This set-up reinforces the vulnerabilities of groups without resources. Lacking such economic resources is often combined with lacking the social, cultural and symbolic capital necessary to acquire the economic assets. Thus, the LEADER concept fails to address the preconditions of participation of the poorest.

References

Alsózsolca Város Önkormányzatának *Integrált Városfejlesztési Stratégiája Szociális célú városrehabilitáció ÉMOP-3.1.1-12*, [*Integrated City Development Strategy for Social purpose rehabilitation for Alsózsolca Municipality*], 2012 [Online]. Available at: <http://www.alsozsolca.hu/letoltes/519/ivs.pdf> [Accessed February 2013].

Asztalos Morell, I., 1999. *Emancipations Dead-End Roads? Studies in the formation and development of the Hungarian Model for Agriculture and Gender (1956–1989)*. Uppsala: Acta Uppsaliensis, 46.

B, Gy., 2013. *Demonstráció a megbékélés jegyében Alsózsolcán* [*Demonstration for reconciliation in Alsózsolca*] [Online]. Available at: <http://www.boon.hu/demonstracio-a-megbekeles-jegyeben-alsozsolcan/2335762> [Accessed 30 June 2014].

Bierbaum, R., 2011. Climate governance and development. In: Ansohn, A. and Pleskovic, B., eds. *Climate governance and development: Berlin Workshop Series 2010*. Washington, DC: The World Bank, pp. 11–26.

Bourdieu, P., 1986. The forms of capital. In: Richardson J.C., ed. *Handbook of Theory and Research for the Sociology of Education*. New York: Greenwood Press, pp. 241–8.

Buzar, S., 2007. *Energy poverty in Eastern Europe*. Aldershot: Ashgate.

Casillas, C. and Kammen D.M., 2010. The energy-poverty-climate nexus. *Science*, 330, November, pp. 1,181–2.

Cecelski, E., 2000. *Enabling equitable access to rural electrification: Current thinking and major activities in energy, poverty and gender*. Washington, DC: Asia Alternative Energy Unit, The World Bank.

Csereháti Roma Önsegítö Közhasznú Egyesület, 2010. *BÜKK-MAK LEADER HACS területén* élő *roma lakosság helyzetének komplex feltárása* [*A complex analysis of the Roma population living in the territory belonging to BÜKK-MAK LEADER HACS*], 2008 manuscript.

Csongor, A., Lukács, G.R. and Higgins, N., 2003. *Labour Market Programmes for the Roma in Hungary Hungarian Foundation for Self-Reliance*. Budapest: ILO Sub-regional Office for Central and Eastern Europe, and UNDP.

Curtin, C.T. and Varley, A., 1997. Take your partners and face the music: the state, community groups and are-based partnerships in rural Ireland. In: Brennan, P., ed. *Securalisation and the state in Ireland.* Caen: Caen University Press, pp. 119–46.

Dupcsik, Cs., 2005. A magyarországi cigány-kutatások cigányképe [Gypsy image of research on gypsies in Hungary]. In: Neményi M. and Szalai J., eds. *Kisebbségek kisebbsége. A magyarországi cigányok emberi* és *politikai jogai* [*The minorirty of minorities. The human and political rights of gypsies in Hungary*]. Budapest: Új Mandátum, pp. 255–82.

Durst, I., 2008. 'Bárók', patrónusok versus 'komák' – eltérö fejlödési utak az aprófalvakban [Barons and patrons versus friends: diverging developmental paths in small villages]. In: Váradi M.M., ed. *Kistelepülések lépéskényszerben* [*Small communities in need to move*]. Budapest: Ùj mandátum Kiadó, pp. 232–81.

Emigh, R.J., Fodor, È. and Szelényi, I., 2001. The Racialization and feminization of poverty?. In: Emigh, R.J. and Szelényi, I., eds. *Poverty, Ethnicity, and Gender in Eastern Europe During the Market Transition.* Westport, Connecticut, London: Prager, pp. 1–32.

Hamar, A., 2010. A csépai szociális földprogram [The social land-programme of Csépa]. In: Váradi, M., ed. *Esélyteremtök* [*Chance-creators*]. Budapest: MTA VITA Alapítvány, pp. 112–31.

Haney, L., 2002. *Inventing the Needy: Gender and Politics of Welfare in Hungary*. Berkeley: University of California Press.

Hegedüs, M., 2007. Energiapolitika: as EU és Magyarország [Energy policy: the EU and Hungary]. *Az Elemzö*, 2007(1), pp. 65–78.

Herrero, S.T. and Ürge-Vorsatz, D., 2012. Trapped in the heat: A post-communist type of fuel poverty. *Energy Policy*, 49(1), pp. 60–68.

Járosi, M., 2007. Magyar energiapolitika lehetőségei és követelményei [The possibilities and demands of Hungarian energy policy]. *Nemzeti* Èrdek, 2007(1), pp. 1–20.

Kaderják, P., Mezősi, A., Paizs, L. and Szolnoki, P., 2010. *Energiapolitikai ajánlások*. Budapest: Regionális, Energiagazdasági Kutatóközpont.

Kajati, Gy., 2011. The geographical environment of the Hungarian electricity industry. In: Kozma, G., ed. *New results of cross-border co-operation.* Debrecen: DIDAKT Kft, pp.103–10.

Kay, R., 2010. Social security, care and the 'withdrawing state' in rural Russia. In: Jäppinen, M., Kulmala, N. and Saarinen, A., eds. *Gazing at Welfare, Gender and Agency in Post-socialist Countries*. Newcastle upon Tyne: Cambridge Scholars, pp. 145–68.

Kearney, B., Boyle, G.E. and Walsh, J., 1994. *EU LEADER I initiative in Ireland: evaluation and recommendations*. Dublin: Department of Agriculture, Food and Forestry.

Kemény, I., 1976. *Beszámoló a magyarországi cigányok helyzetével foglalkozó 1971-ben végzett kutatásról* [*Account on the 1971 research on the conditions of Gypsies in Hungary*]. Budapest: MTA Szociológiai Intézet.

Kemény, I., Jankay, B. and Lengyel, G., 2004. *A magyarországi cigányság, 1971–2003*. [*Gypsies in Hungary, 1971–2003*]. Budapest: Gondolat.

Kotics, J., 2012. A csermelyi 'Biomassza' project és a helyi társadalom: A cigány-magyar együttélés konfliktusai és azok hatása a project megvalósitására' [*Biomas project in Csermely and the local community: Conflicts of the Roma-Hungarian cohabitation and its impact on the realisation of the project*]. *Szellem* és *Tudomány*, 2012(1), pp. 38–69.

Kovách, I., 2000. LEADER, a New Social Order, and the Central- and East-European Countries. *Sociologia Ruralis*, 40(2), pp. 181–90.

———, 2008. Kistelepülések lépéskényszerben, in Váradi, M., ed. *Kistelepülések lépéskényszerben* [*Small communities, under pressure*]. Budapest: Új Mandátum Könyvkiadó, pp. 7–28.

———, 2010. *A jelenkori magyar vidéki társadalom szerkezeti* és *hatalmi változásai*. MTA doktori értekezés [Hungarian Academy of Sciences thesis]. Budapest: MTA Hungarian Academy of Sciences.

———, 2012. *Access of Roma Communities to EU Funds in Bulgaria and Hungary*. Unpublished paper presented at the International Rural Sociological Association Conference, Lissabon.

Kovách, I. and Kucerová, E., 2006. The Project Class in Central Europe: The Czech and Hungarian Cases. *Sociologia Ruralis*, 46(1), pp. 3–21.

Kovács, I. and Zubor, Z., 2013. *'Szétesett a társadalom' a magára hagyott Alsózsolcán* [*'Disintegrated society' the abandoned Alsózsolca*] [Online]. Available at: <http://hvg.hu/itthon/20130906_Alsozsolca_riport> [Accessed 30 July 2014].

Krishna, A., 2011. *One illness away: Why people become poor and how they escape poverty?* Oxford: Oxford University Press.

KSH, 2010. *Magyarország 1989–2009, A változások tükrében* [*Hungary 1989–2009, In the context of changes*]. Budapest: KSH.

Ladányi, J. and Szelényi, I., 2004. *A kirekesztettség változó formái*. Budapest: Napvilág.

———, 2002. Cigányok és szegények Magyarországon, Romániában és Bulgáriában. Gypsies and poor in Hungary, Romania and Bulgaria), *Szociológiai Szemle*, 2002(4), pp. 72–95.

Lázár, I., 2010. *Vidékfejlesztési stratégiák* és *a fenntarthatóság*. Unpublished Study presented at Corvinus Egyetem, Budapest.

Macken-Walsh, I. and Curtin, C., 2012. Governance and rural development; The case of the rural partnership programme (RPP) in Post-socialist Lithuania. *Sociologia Ruralis*, 2012(53), pp. 246–64.

Marsden, T., 2008. *Sustainable Communities: New spaces for planning, participation and engagement*. Oxford: Elsevier.

Molnár, E. and Dupcsik, Cs., 2008. *Country Report on Education: Hungary*. Budapest: Edumigrom, CEU.

Molnár, E. and Schafft, K.A., 2003a. A helyi roma/cigány kisebbségi önkormányzatok tevékenysége és céljai Magyarországon 2000–2001-ben [The activities and goals of local Roma/gypsy minority self-governments in Hungary in 2000–2001]. *Szociológiai Szemle*, 2003(1), pp. 79–99.

———, 2003b. Social exclusion, ethnic political mobilization, and Roma minority self-governance in Hungary. *East Central Europe*, 30(1), pp. 53–74.

Martinot, E. and Li, J., 2007. *Powering China's development: the role of renewable energy*. Washington, DC: Worldwatch Institute.

Narayan, D., 2002. *Empowerment and Poverty Reduction: A Sourcebook*. Washington, DC: World Bank.

O'Hara, P., 1998. *Action on the ground: models of rural development practice*. Galway: Irish Rural Link.

Polányi, K., 1976. *Az archaikus társadalom* és *gazdasági szemlélet* [*Archaic society and economic view*]. Budapest: Gondolat.

Putnam, R.D., 2000. *Bowling alone: The Collapse and Revival of the American Community*. New York: Simon Schuster.

Rose, R., 1998. *Getting Things Done in an Anti-Modern Society: Social Capital Networks in Russia*. Washington, DC: World Bank Social Capital Initiative, Paper No. 6.

Sätre, A., 2012. Gendered experiences in entrepreneurship, family and social activities in Russia. In: Carlbäck, H. et al., eds. *And they lived happily ever after Norms and everyday practices of family and parenthood in Russia and Easter Europe*. Budapest and New York: CEU Press, pp. 297–318.

Sen, A., 1984. *Resources, Values and Development*. Oxford: Basil Blackwell Publ. Ltd.

Schwartz, Gy., 2012. Ethnicizing poverty through social security provision in rural Hungary. *Journal of Rural Studies*, 28(2), pp. 99–107.

Shortall, S. and Shucksmith, M., 1998. Integrated rural development: issues arising from the Scottish experience. *European Planning Studies*, 6(1) pp. 73–88.

Shucksmith, M., 2000. Endegeneous Development, Social Capital and Social Inclusion: Perspectives from LEADER in the UK. *Sociologia Ruralis*, 40(2), pp. 208–18.

Shucksmith, M. and Chapman, P., 1998. Rural Development and Social Exclusion. *Sociologia Ruralis*, 38(2), pp. 225–42.

Söderbaum, P., 2011. *Bortom BNP, Nationalekonomi och företagsekonomi för hållbar utveckling* [*Beyond BNP, Economics and business administration for sustainable Development*]. Lund: Studentlitteratur.

Svendsen, G.L.H. and Svendsen, G.T., 2004. *The Creation and Destruction of Social Capital, Entrepreneurship, Co-operative Movements and Institutions*. Cheltenham: Edward Elgar.

Szabó-Tóth, L., 2012. Csermely esete a biomasszával. Csernely energiagazdálkodási modelljének kidolgozását megalapozó lakossági felmérés eredményeinek bemutatása' [The case of Csermely with the biomass. Presentation of the community survey serving to elaborate Csermely's the model for energy economy]. *Szellem* és *Tudomány*, 2012(1), pp. 38–69.

Szalai, J., 2005. A jóléti fogda' [The welfare prison]. In: Neményi, M. and Szalai, J., eds. *Kisebbségek kisebbsége. A magyarországi cigányok emberi* és *politikai jogai* [*Minority of minorities. Human and political rights of the Hungarian Roma*]. Budapest: Új Mandátum Könyvkiadó, pp. 43–93.

———, 2007. *Nincs két ország? Társadalmi küzdelmek az* állami *túlelosztásért a rendszerváltás utáni Magyarországon* [*Are there not two countries? Social struggles for state overdistribution in post-socialist Hungary*]. Budapest: Osiris.

Szoboszlai, Zs., 2004. Szegénység, marginalizáció, szegregáció, [Poverty, marginalisation and segregation]. *Tér* és *Társadalom*, 18(3), pp. 25–42.

Swain, N., 1994.Transition from collective to family farming in post-communistic Central Europe. *Eastern European Countryside*, pp. 17–30.

Szuhay, P., 2005. (Utó) parasztosodó törekvések a szendröládi romák körében [(Post)peasantification aspirations among the Roma in Szendröläd]. In: Schwarcz, Gy., Szarvas, Zs. and Szilágyi, M., eds. *Utóparaszti hagyományok* és *modernizációs törekvések a magyar vidéken* [*Post-peasant traditions and modernisation approaches in rural Hungary*]. Budapest: MTA Néprajzkutató Intézet, pp. 59–74.

Thelen, T., Dorondel, S. Szöke, A. and Vetters, L., 2012. 'The sleep has been rubbed from their eyes': social citizenship and the reproduction of local hierarchies in rural Hungary and Romania. *Citizenship Studies*, 15(3–4), pp. 513–27.

Tódor, J., 2012. *Miért lőttek Rontó Kirsztiánra* [*Why did they shoot Krisztián Rontó*] [Online]. Available at: <http://www.commmunity.eu/2012/06/14/miert-lottek-ronto-krisztianra/> [Accessed 13 April 2014].

UNDP, 2005. *Faces of poverty, faces of hope. Vulnerability profiles for decade of Roma inclusion countries* [Online]. Available at: <http://roma.undp.sk/> [Accessed 12 April 2014].

Vajda, R. and Dupcsik, Cs., 2008. *Country Report on Ethnic Relations: Hungary*. Budapest: Edumigrom, CEU.

Váradi, M., 2008. *Kistelepülések lépéskényszerben* [*Small settlements in need to act*]. Budapest: Ùj Mandátum.

World Bank, IBRF, 2000. *Making transition work for everyone: poverty and inequality in Europe and Central Asia*. Washington, DC: World Bank.

Interviews

Interview 1 (2012) Attila Kardos, County Vajda, proposed waste project leader.
Interview 2 (2012) László Rózsa, Romani Minority Self-government leader.
Interview 3 (2013) Mayor of municipality A, LAG member.

Interview 4 (2012) Antal Láng, LAG manager.
Interview 5 (2013) Albert Balogh, capital strong Romani entrepreneur.
Interview 6 (2013) Jobbik (right-wing radical party) politician.
Interview 7 (2012) Mayor of municipality B, LAG member.
Interview 8 (2013) Mayor of municipality D, LAG member.

Chapter 12

Developing or Creating Instability? Development Management, Scale and Representativeness in Tunisia

Aude-Annabelle Canesse

Introduction

Since the Independence of Tunisia (1956), development has been a constituent of the political discourse and, in terms of public policies, has been based on large-scale national programmes that constitute a modality of development such as the five year plan or, in the days of Ben Ali, the presidential decisions.[1] Because of its significance for the territory, early national programmes focused on the rural sector (Rural Development Programme; Integrated Rural Development Programmes, etc.). However, whereas territorial inequalities and opposition between the rural and urban sector had been highly underlined in academic literature (Baduel 1985; Belhedi 1989, 1992; Sethom 1992), rural development policies seem to have had limited impact. Indeed, in 1978, 1984 or considering the Arab Spring in 2010–2011, the Tunisian uprisings started in rural areas and illustrate the development failures. The aim of this chapter is to analyse development policies, institutions and instruments that enhance local rural development and participation in Tunisia during the last decades.[2] Whereas the country is often considered under the light of development studies as part of the 'South', it addresses development institutions and instruments and thus opens the door to a comparison with research on Northern countries (Sjöblom et al. 2006; Andersson 2009), especially through development management. Besides, three years after the so-called 'Revolution' the elements presented below help at understanding the development components, especially the management, the scale and the (lack of) representativeness, that determined the last but not least Tunisian uprising and that are still at stake in 2014, being structural factors of instability.

Generally speaking, projectification programme-based approach to public policy contains a paradox. In their very nature, programmes represent a process of

1 However, presidential decisions belong to the measures identified by regional administrative services during the plan elaboration and may finally not be chosen by central departments.

2 The author is grateful to Brian Pasco for useful comments.

inclusion and exclusion. A project's content is neutral, technical and organisational, according to performative objectives (Craig and Porter 1997) and is not political. It associates ideas, suggestions, resources, men and places, while leaning on criteria, procedures, documents, and techniques. The main goal is timeless, and based on ideals whereas the specific objectives are observable, objectively defined, and leaning on project mechanisms that include the target populations. The rationalisation of objectives is based on the logical framework approach that can be considered as a public policy tool. This proposes quantifiable objectives: a general goal, specific objectives, means and indicators allowing evaluation of whether the objectives are reached. In parallel, this translation of public policy, into the 'LogFrame' rationalisation, constitutes a legitimising discourse. Giovalucchi and Olivier de Sardan (2009) identify in the logical framework the political discourse invariants: the transparency of reality through a presentation under quantifiable measures; the legitimisation of financial backers which justify their choices; the control of social phenomena by the logical framework which is supposed to integrate all the dimensions and to prescribe activities according to a 'logical' procedure; the designation of the stakeholders, 'beneficiaries' or 'targeted population' designated as an undifferentiated whole, while denying local dynamics and conflicts and more broadly all the obstacles that could be met (Rondinelli 1976). Projects identify beneficiaries, needs and circumscribe geographic zones. As such, they lead to a process of inclusion and exclusion, including those and only those who correspond to these categories or those who are able to adapt themselves according these criteria, and who are at the right place, at the right moment (i.e. who have the information). As a consequence, the expression of population is limited; it has to express its needs according to the framework's assumptions.

However, this does not mean that programmes and projects are useless: as instruments, 'technical plans with a generic intent, symbolizing a concrete relationship between politics and society, and supported by an idea of regulation' (Lascoumes and Le Galès 2007),[3] they illustrate the relationship between state and society. In the Tunisian case, they are a method to frame territory and population and to reproduce authoritarian political practices and institutions. To make this argument, this chapter tackles objectives and institutions, 'the routines, procedures, conventions, roles, strategies, organizational forms, and technologies around which political activities are constructed' (March and Olsen 1989), instruments and the role of actors involved in local development. This chapter is based on a long fieldwork and specialisation in Tunisia – an 18-month continuous stay in 2005–2007 and additional longer and shorter stays since 2004 combining academic and applied research on international development programmes – as well as a wide range of materials: evaluation reports, interviews made with evaluators and actors involved in development (Ministry of Agriculture and Hydraulics Resources, Ministry of Interior and Local Development, Ministry of Planning, NGO,

3 '*Dispositifs techniques à vocation générique porteurs d'une conception concrète du rapport politique/société et soutenus par une conception de la régulation*'.

international actors and 'beneficiaries'). The areas visited and chosen during the formulation were previously included in national programmes (Integrated Rural Development Programme; Integrated Agricultural Programme). Thus the chapter also gives concrete information of programmes' impact on development.

To understand the context, the first part of the chapter briefly reviews the Tunisian institutions involved in development; then it addresses the first programme in the rural sector, the Rural Development Programme (RDP, initially called Regional Development and Rural Animation Programme), presenting its objectives, its organisation and its socioeconomic and political impacts. These will remain similar decade after decade despite other national programmes (Integrated Rural Development Programmes) or international ones. Lastly the chapter focuses on recent institutions in the rural sector (Agricultural Development Group, ADG), and their implementation within the framework of a recent international programme. With regards to former natural resource management groups, ADG have diversified activities: they are in charge of both natural resource management and of local development, taking into consideration technical and social dimensions (natural resource conservation, etc.; participation). In addition, they refer to the withdrawal of state, as its initial competencies are transferred to private sector and civil society, in both local decision-making and financial aspects of maintenance.

Tunisian Institutions Involved in Development: Framing Local Participation

The political discourse emphasises the importance of participation which would formally take shape through decentralisation and the creation of councils at each administrative and territorial level, and more widely in local development. Based on a hierarchical distribution of political power, two categories of institutions can be distinguished that correspond to the 'administrative authorities' and the 'political authorities', as called in the interviews. On one hand an institutional system of governance (Canesse 2010, 2013) is composed of the rural council, the local development council and the regional council; on the other hand the territorial representations of the Democratic Constitutional Rally (DCR) consists of the presidential party that is situated at the crossroad between administration and political party. In rural areas, the DCR local cells are all the more important since there are no elected representative bodies: thus cells constitute places where (part of the) population can express wishes, and where tensions appear between families or interests over access to political, social and economic resources. Besides, in these areas, this is the only party.

An Institutional System of Governance in the Rural Area

The Rural Council (1989) is an advisory body in rural areas with no municipality and is consulted for economic, social, cultural and educational issues, transmitting population needs, proposing solutions or being involved in programmes' implementation (Figure 12.1). Its members are designated by the Ministry of Interior and Local Development (MILD), considering their designation by the *omda*,[4] the Ministry of Interior's representative at local level.

The Local Development Council (LDC) was created in 1994 at the delegation level, between local and regional levels. As is the case with the Rural Council, this is an advisory council involved in local development programmes and projects, regional development planning, and execution of environmental protection programme and natural resource management (rationalisation, preservation). The LDC is composed of elected members (municipalities and arrondissement presidents) and designated members (Rural Council's presidents; *omdas*, administration regional services' representatives). They are under the authority of the governor who is the Ministry of Interior's representative at the regional level.

Year 1989 was a turning point. Besides the creation of Rural Councils, the Regional Councils replaced the Councils of Governorate. The Regional Council is in charge of socio-economic programmes and land settlement[5] plans, scheduling and formulating these plans, as well as examining urban plans and implementing national policy. Concerning the state's programmes, the Regional Council has advisory role and also coordinates national, regional and municipal programmes. This council is composed of some designated members (governor, governorate general secretary, presidents of Rural Councils) and some elected ones (deputies, presidents of municipalities). Contrary to the *omdas* and the delegates, the governor has a political trajectory in the Democratic Constitutional Rally and has broad competencies representing the state and applying national directives. He acts as the president of regional council and also of rural council.

The Democratic Constitutional Rally, Between Political Party and Administration

The electoral process at a micro level, and more broadly in rural areas, is at stake as they have no local elected representatives – except those of the former single party, the Democratic Constitutional Rally, the local political 'cells', which have several functions.[6] They are an intermediate level between centre and periphery, to use the concepts by Grémion (1976), analysing French decentralisation that enables the representation of local interests and their particularisms to the centre. Their sites respond to the necessity of watching and of establishing power relationships and

4 The *chef de secteur* is the Ministry of the Interior and Local Development's representative at the rural level. He is in charge of supporting administration services. He is also a judicial police and a registry civil officer. In practice, he represents both the administration and the notables (Hénia 2006).

5 Outside of areas of municipalities.

6 In 2004, the Tunisian territory had 7,500 local political cells, the sites of which overlap administrative territories, and 2,200 professional cells, the members of which were elected.

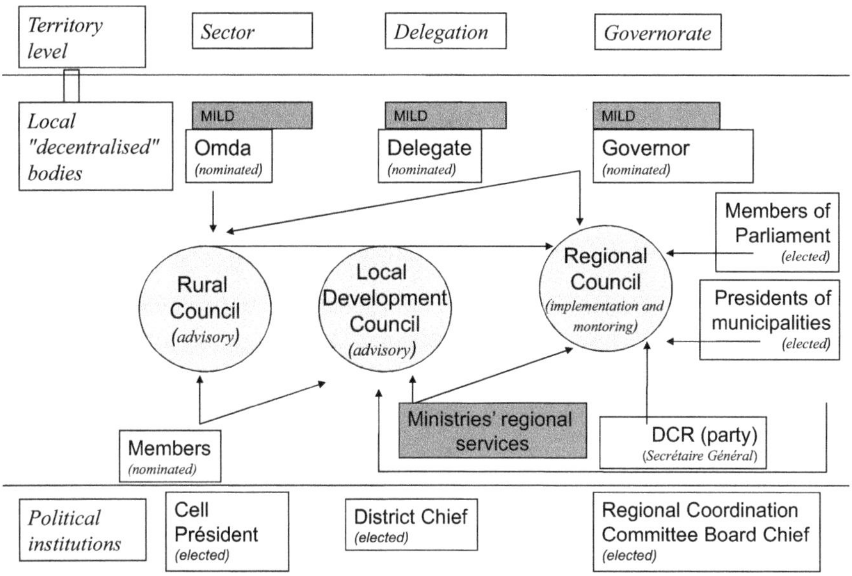

Figure 12.1 Local Bodies in Tunisian Rural Areas

Source: Aude-Annabelle Canesse, 2010.

they are used by elites as a place of 'mediation between local interests and politico-administrative institutions of the state' (Hibou 2006).[7] At the local and micro-local level, the DCR cells reflect local conflicts as well, as they combine local interests and conflicts. The central power can support or break up those tensions to improve public policies or to impose measures decided at the central level.

Besides the spread of ideological discourse, to some extent the political cells represent a parallel administration as underlined in the interviews. For individuals, they make administration procedures easier and faster; adhering to DCR and getting involved in cells or in 'political associations'[8] also boils down to join a network to quickly obtain a job or even microcredits. Thereby, benefits are more financial than ideological, for youth in particular, whereas elderly people associate DCR with the former parties, *Destour* and *Neo-Destour* linked to the fight for Independence. On the contrary, not being a member, especially in rural areas leads to being excluded. Moreover, DCR political cells are involved in public policy implementation, identifying 'shadow areas' targeted by aid programmes, or being involved in professional consultations and follow up committees, such as the ones concerning natural resource management groups.

7 '*médiation* entre les intérêts locaux et les institutions politico-administratives de l'Etat'.

8 A way to call student associations for instance (Interviews 2008).

Thus in Tunisia, and despite the political discourse, the recognition of the local level is limited; whereas the Rural Council and the Local Development Council officially constitute bodies representing population and transmitting information from local levels to central level, they do not meet criteria of representation as most of their member are designated and not elected. The governor, the Ministry of Interior and Local Development's representative, remains central. Added to the cells of the Democratic Constitutional Rally, these 'administrative and political authorities' involved in participation and more widely in local development and public policy implementation constitute a *maillage*, meant as organising a group or persons within an administrative or political network that facilitates governing and has a function of social project framing (Maurel 1984). They enable the transmission of information and the control of population and territory.

The Local Development Programmes as an Illustration of the Permanence of Institutions and Objectives (1973–2006)

Since 1973, national programmes have been implemented in rural areas, notably the Rural Development Programme followed by the Integrated Rural Development Programme, together covering three decades of development policies. They help to demonstrate permanency in practices and the development failures, and to understand the recent context, including the territorial and social inequalities that led to the Arab Spring. Indeed the permanency of socioeconomic issues means problems have remained stable and partially unsolved, leaving rural population disadvantaged despite coastal development and good national macroeconomic results. Besides, it focused attention on the high institutional dimensions, limiting the population's involvement in the decision-making process and the importance of harnessing funding.

The Rural Development Programme (1973–1982)

The development model was based on cooperatives and polarity before the 1960s, when it turned out to have failed (see Perroux, 1964).[9] A decade later, dramatic territorial disparities lead to rural exodus as well as increasing socio-economic problems and pauperisation in urban areas. As a consequence, the Regional Development and Rural Animation Programme (also called Rural Development Programme, RDP)[10] was implemented and aimed at reducing the gap between rural and urban areas and between the regions themselves, focusing on rural

9 Political choice of development are presented in detail by Canesse 2014.

10 The way to call the programme is interesting. Indeed its name changed depending on several decrees and institutions, emphasising or hiding some aims. Finally from 1977 onwards the rural animation dimension was outshined, to the profit of rural development only.

activities, on employment, and on livelihood conditions. Therefore four categories of population were targeted: the poorest, the unemployed, the youth, and some population that were not the most disadvantaged. These measures aimed also at settling rural population and, from 1980, at implementing the five year plan in the governorate. From this time, the RDP has appeared to be a complement to the national measures, financing small and medium local and regional projects that could not be supported by national measures. The emphasis was no more on peasantry but on 'small and middle scaled projects' and on 'private promoters', revealing the primacy given to economic dimension and to the market.

In terms of management, the RDP reproduced the five year plan elaboration and implementation. Based on regional and national levels, the general objectives were decided in Tunis; at the regional and at the delegation levels, under the aegis of the governor, actors belonging to administration or to partisan organisations[11] identified measures they transmitted after to the national commissions in Tunis, once being validated by the technical commission. At the national level, the Direction of the Regional Planning presented them to the National Commission chaired by the Prime Minister before being discussed at the National Assembly. Once the projects were accepted, the governor committed the credits granted by the Ministry of Plan, under control of the Ministry of Finances; the RDP service[12] was in charge of programme's execution and animation, assuring the connection between local, regional and national levels whereas the implementation returned to the technical services in the governorates.

However, local and regional needs had to be expressed in the terms of national policies' orientations and in addition, in this strong institutional environment it appeared that people's involvement was limited, favouring clientelism and short-termism. Indeed, participatory and decision-making limits were important due to the absence of monitoring of actions and beneficiaries as well as of maintenance structures, and furthermore, the overlapping of sub-programmes assured by separate services, and the elaboration under the pressure of local population's dire needs, especially as funds were assured for one year, that is to say a very short time, that led to the funds to be spread thinly and to partial failure. The practise also suffered from the absence of representative structures in which the population could be involved, whereas the political structures and imperatives in choosing beneficiaries, actions and localities implied negative reactions from both populations and technical services (République tunisienne, Ministère du Plan, Commissariat général au Développement régional 1983, 4–5). As such the RDP activities boiled down to a mode of local political management enabling the framing of the population. At the local level, the single-party cells (*Neo-Destour* at the period), the delegate and the *omda* played an important role in

11 At this time, the political system is based on a single party. In other words, unions and professional associations are embedded in it.

12 The formerly named 'Office of Rural Animation' becomes a service in 1977, the staff and the material of which is assumed by the Ministry of Interior.

programme's implementation by choosing beneficiaries, equipment and spreading crucial information about the RDP, favouring parental and/or partisan closeness. Besides, almost 30 per cent of the beneficiaries declared they never had requested any actions from the programme.[13] The funds also enabled equipment, even the construction, of political cells, helping then the single party to get established at the local level. In the field of housing, the actions illustrated the inadequacy with population needs; several housing remained vacant and 65 per cent of the beneficiaries declared themselves dissatisfied due to their site and their conception.

A New Model of Development and Management? The Integrated Rural Development Programme

Following the Rural Development Programme, the Integrated Rural Development Programme (IRDP) was implemented and put the light on integrated development in its several dimensions (in both the area and the parcel, and linking productive actions to improvement of living conditions). These actions addressed agriculture, small trades, services and infrastructures, and the objectives remained very close to the RDP: production increase, income improvement, employment creation, integration of productive interventions linked to livelihood conditions and standard of living improvement, beneficiaries' participation improvement in particular in project identification, execution and financing. With regard to the RDP, the zones were widened and every programme covered one or two territorial sectors; at higher level, the Ministry of Agriculture elaborated similar programmes with regional vocation.

In terms of implementation, the Regional Development General Commission was the main actor, reporting to the financial backers and carrying out actions' planning and coordination, or handling the preparation of technical studies by resorting to engineering consulting firms. The IRDP followed a three step implementation: first the identification of territories[14] and actions by way of socioeconomic surveys to specify actions, needs and choices made by participants. Second, the importance of development management was emphasised; project managers had a pre-eminent role, being in charge of socioeconomic inquiries, sensitisation and information meetings organised in association with regional and local authorities. However, despite some changes in the management, the IRDP encountered organisational and participative limits, notably considering the areas designation or the management, considering the formal administrative process that remained characterised by slowness (i.e. for the call for tenders; request for

13 Concerning the beneficiaries, it is interesting that no list was made available, at least for external persons of the project.

14 Localisation, surface, occupation and types of activities, natural potentialities, the 'main characteristics of the population', infrastructures and main activities.

beneficiaries)[15] or the programme management. According to the appraisal report, the IRDP managers did not have 'the adequate profile' and their availability was put into question.[16] Lastly, the (not systematic) feasibility study preceded the socioeconomic study of the results: actions were chosen in the zones without knowing if the socioeconomic conditions were adapted; the actions could be modified during the process, and the participation could be 'guided' to obtain the envisaged results. Therefore, the participation could still be considered as 'formal'.

The participation in a project design mission in Tozeur (2006–2007) enabled the visiting of several IRDP sites created between 1986 and 2003, and made it possible to identify the inadequacy between the programme's actions and the population's needs. In the 2000s, some sites still lacked infrastructures of protection and houses, and the most recent date palms had not reached yet the full production and could not benefit from 'oasis effect' enabling intercropping system. Contrary to traditional oases, the new palm groves mentioned here as irrigated perimeters did not benefit from traditional local institutions spreading knowledge and coordinating beneficiaries' tasks. These beneficiaries were able to choose between subsidies and credits with the Agricultural National Bank (in particular for a female camel and 10 ewes); however the subsidy, subordinated to the refund of the credit for animals, did not benefit all the farmers. Besides, few beneficiaries with small parcels had sufficient financial capital and means, sometimes even knowledge, to start a new lifestyle in an oasis environment. In spite of the creation of basic infrastructures around the new irrigated perimeters,[17] it remained difficult for the new farmers (or agro-breeders) to start productive initiatives in these areas, sometimes far from cities. In addition, in many irrigated perimeters the housing environment remained unsuitable for populations living in towns; here, the limits were identical to the Rural Development Programme.[18]

---

15 The request was elaborated by the project manager and signed by the beneficiary; then a regional committee established the list of beneficiaries and transmitted it to the governor. After the governor's signature, the General Secretary of the governorate requested for loans.

16 Indeed, several project managers were foresters and were in charge of several IRDP zones. They could also have activities within the Ministry of Agriculture departments in the governorate.

17 Paths, plantations, irrigation system, single room houses, systems of protection of the perimeter, sometimes health centre, a school, etc.

18 The case of Ben Guech, an irrigated perimeter situated at the border between Tunisia and Algeria, is interesting. It is recognised by the administration as a case of failure: 19 of 48 beneficiaries abandoned their land plot, sometimes even before the palm grove installation. Several factors are put forward: criteria of land plot distribution, the isolation of the irrigated perimeter on the Tunisian territory, the distance from Nefta and Tozeur. Nonetheless, few hundred meters from Ben Guech, an eponym irrigated perimeter in the Algerian territory, well-functioning before the Tunisian unity, could be noticed.

In the 2000s, New International Programmes but the Same Issues

International programmes constitute another modality of local development. The examples addressed here ($12 million international development programme in Kairouan, Siliana and Zaghouan, 1999–2006; project formulation in Tozeur, 2006–2007) illustrate the multiple dimensions of development issues in Tunisia in the more recent years – added to change and permanency. Whereas development management seems to have been taken into account, it appears that development issues still remain similar, showing that previous development actions failed. Widely, these four governorates are characterised by unemployment combined with demographic pressure, rural exodus, similar houses, opposition between a majority of small exploitations and a few big exploitations and the absence of land title. They also illustrate political dimensions in development through the control of territory and population.

Changing Development Management?

In recent years, programme implementation by experts, apparently from the non-state sector, is favoured instead of implementation held by public administration. Authors distinguish different categories of experts and field of expertise (see Dumoulin, La Branche, Robert and Warin 2004; Lequesne and Rivaud 2001) but the Tunisian fieldwork revealed a new category, the experts of delegation: temporary consultants from private firms, and animators and programmes' coordinators recruited by the international organisation designated as the executive agency (Canesse 2012). This transfer of activities from administration to the non-state sector is justified by functional arguments related to a better management, neutrality and efficiency. Consultants, animators and coordinators are considered to be more neutral and more efficient, while not belonging to administration, and the two last categories would be full time devoted to the project. As a consequence, at the first glance it seems that this kind of recruitment fills the gap observed during the former Tunisian programmes. Yet, whereas their recruitment changes and they officially depend on the non-state sector, their career path and their training are close to administration; some of them even overlap a double institutional position (national administration/international organisation in the case of coordinators) and some of them can even be considered to belong to administrative elites. As a consequence, their practices include political issues leading to the reproduction of the authoritarian system practices.

However, if no real change can be observed in the development management, the public administration is directly impacted; it is internally reconfigured and its activities are externalised (Canesse 2012, 2014) to these actors who constitute a 'project class' (Kovách and Kučerova 2009) gravitating around development programmes.

Targeting Underdeveloped Rural Areas

In the governorates of Kairouan and Zaghouan, the majority of the parcels are less than five hectares. In Kairouan, a governorate suffering from farmers' underemployment and unemployment, most of the local labour moved to big cities to work on construction sites. In the governorate of Siliana, the demographic pressure is very strong and as in Kairouan, men exercise extra-agricultural activities in the city. For several enclosed *dechrats* or *douars*,[19] the priority requests remain access to drinking water and rural path. In the governorate of Zaghouan, close to Tunis and often presented as a governorate 'with problems' and even as a governorate 'without solutions' (Interviews 2004, 2006), unemployment and pressure on the resources are stronger due to economic difficulties and population growth. The areas in Zaghouan results from colonial period and opposes large and small exploitations (mainly varying from 1 to 4 ha), the latter attributed to the native inhabitants to settle them and to constitute a reservoir of available labour.[20] The big exploitations, despite production intensification and diversification for some of them, do not absorb the labour. Aside from farming, the local economy leans on administrative interventions within the framework of water and soil conservation construction or on extra-agricultural activities. In Tozeur, three categories of populations are targeted within the project: nomads and settled semi-nomads; nomads and urban population; native from the city, generally young promoters and unemployed young graduates. As written above, some irrigated perimeters created by IRDP between 1986 and 2003 still lack land protection infrastructures or houses and the most recent date palms do not reach yet full production. In addition, some areas, mainly occupied by nomads, have already benefited from development actions (drinking water supply, electrification, rural path or road asphalting). In the delegations of Hazoua and Tameghza, in Tozeur governorate, the new oases represent 599 ha and 549 beneficiaries (most of them were *khammes*,[21] labourers and some of them pastoralist). These reduced land plots were attributed within the framework of IRDP or equivalent programmes of the Ministry of Agriculture.

The Unsolved Question of Land Status

Land status is a recurring constraint in several rural and agricultural activities we have observed. In Kairouan, where the major part of land is collective and state-

19 The *dechrats* and the *douars* gather houses, the inhabitants of which are linked by family relationships.

20 This situation remained after the Independence, sometimes being deteriorated after division following inheritance, whereas large-dimension exploitation were rarely divided and sometimes even expanded.

21 The *khammes* cultivate parcels that they do not own and keep one-fifth of the production.

owned, most of the beneficiaries do not own an official title and have recourse to the right of use. In Siliana, the land use is rather privative for farmers and collective in the hill and forest paths; a third of farmers do not own a title. The situation in Zaghouan governorate is pretty similar for small exploitations, and only five heads of households signed an annual renewable contract.[22] In the irrigated perimeters of Tozeur, most of the land plots are contractually rented.[23] But apart from these programmes, the governorate of Tozeur – and the South adjacent governorates – presents specificities due to the tribes. From the nineteenth century onwards, their leaders participated in a natural resource management committee through which they controlled paths and pasture use. However, their involvement in this advisory committee was only a formality, especially with regards to the question of land expropriation (Hibou 2006). Recently, in the delegation of Tameghza (Tozeur), some members from the Aouled Sidi Abid tribe were settled while maintaining pastoral activity, and hills of Tameghza are considered as their collective lands and not as state-owned. Lastly in El Frid, the whole population is still nomad and land is collective.

Securing the Borders: The Governorate of Tozeur, an Exemplary Governorate

The border is a political process. Far from being a single line on a map, it is the *sine qua none* condition of state existence and its functions are based on a double movement between countries, conquest or compromise (Ratzel 1988). This is what our field trip in Tozeur, especially in El Frid, revealed, as the exchanges with the local state representatives during which the imperatives of population settlement and of border securing were strongly perceptible. El Frid, where the Ministry of Interior held a development project, is a highly controlled area and our visit on the field required the presence of the delegate; on the site, some militaries were present and it was forbidden to film or to take pictures. Thus, development activities (e.g. the creation of new palm groves) are not limited to productive functions. Being located near the border, they are linked to the territory control – and as such have political functions. This link between techniques (i.e. production) and control has two dimensions: on one hand it helps at securing the line of the border, that is to say the state territory; on the other hand, the new palm groves, implemented within the framework of the IRDP for the greater part, enable to limit the '*pass'port cam'ra*' ('passport of the moon') – expressing the way to cross the border under the full moon light.

Lastly, one of the institutions in charge of development in Southern Tunisia enhances this political dimension and the importance of the border closeness.

22 The contract enables them to exploit a 10 hectare forest land plot for 29 TND/ha.

23 There are two types of contracts according to the programme within which these have been signed: a 40- or 25-year lease (IRDP) or a 25-year lease (agricultural programmes).

Contrary to the other Development Offices,[24] the Development Office of R'Jim Maâtoug, located near the Tuniso-Algerian border, has the peculiarity to depend on the Ministry of Defence. While Development Offices are industrial and commercial public bodies, the Development Office of R'Jim Maâtoug is a non-administrative public establishment the decisions of which are not made public. Among its activities, this institution is in charge of a development programme combining also technical and political dimensions. It aims at attributing land plots, to support dates production, and to settle population, nomads in particular.

Strengthening Local Participation in Recent International Development Programmes

Thus, decade after decade, development issues remain consistent – and unsolved – and as such this illustrates limited impacts, even failures, of development policies, despite change in development management, and historically, local development appears to help at controlling population, and territory at enhancing control. This part addresses an international programme in recent years that focuses on strengthening rural participation. Some changes can be underlined, notably in terms of actors; international programmes support the appearance of NGO and the implementation of national participatory plan. In the rural sector these are Agricultural Development Groups (ADG), the implementation of which reveals a multiplicity of interactions between actors with unequal resources around a crucial issue, the territory control. Three categories appeared, enhanced by the national consultant's methodology: the 'development institutions' (as called in interviews), either administrative (Ministry of Interior and Local Development; Ministry of Agriculture and Hydraulics Resources) or political (the Democratic Constitutional Rally), and the notables.

The Recourse to the NGO, Pledge of Effectiveness and of Neutrality?

As the experts of delegation, NGOs belong to the project class and are actors enabling state's activities externalisation and in the Ben Ali era, they are a major stakeholder. Having recourse to international discourse and perspective, they are considered as actors of public policies and are called Non-Governmental Organisations – whereas they are not. Indeed, as their headquarters are located in Tunisia, they are associations and more dramatically submitted to the political

24 The four Development Offices cover several governorates and their follow-up assumed by the Regional Development General Commission (RDGC, Ministry of Development and International Cooperation) which is also in charge of coastal regions programmes' implementation.

context.[25] Few of them were independent, that is to say not partisan and sometimes called 'Governmental Non-Governmental Organisations'; they were included in the framing of society, a process that characterises the Tunisian authoritarian system. In the rural sector, associations were few and recent: in the field of farmers' training, associations were 'training themselves for trainings' (Interview 2004). Thus, and indirectly, local development international programmes support associations' creation and activities.

However programmes' activities faced obstacles that can be explained by the overlapping of procedures and categorisation between associations and the programme, which led to an additional process of inclusion/exclusion and to slowness and gap in beneficiaries' activities. The sector of microcredit illustrates this confrontation. In terms of procedures, associations granted only farmers who had guarantees, while this was opposed to the very nature of microcredits and to the convention signed with the programme. The responsibility was attributed to the Tunisian Bank of Solidarity from which funds were transferred to the associations and which imposed 80 per cent of recovery rate; associations had to reimburse in case a beneficiary could not. But associations also redefined beneficiaries according their own categories and procedures. As an animator underlined 'The population who benefits from microcredits, it is not the population from here' (i.e. the population concerned by the project) and 'We have a convention that stipulates things, microcredits, trainings. And that, it must be respected' (Interview 2004).

External interventions in local development were not limited to the Tunisian Bank of Solidarity. Granting of microcredits also revealed strong administrative interventions, especially from the Ministry of Interior representative at a local level. Indeed, the *omda* identified beneficiaries and had a decisive action concerning associations' activities as an animator emphasised during a development programme's workshop: 'the politicisation of microcredits mustn't be forgotten: the delegate and the *omda* give their opinion at the end. If the person is a practicing Muslim, the delegate and the *omda* remove him, even if he matches the criteria: 'Not him he is not participating to the 26.26',[26] 'Not him, he is a practicing Muslim' (Discussion during a programme's meeting).

Implementation of National Plan and Local Strategies

Agricultural Development Groups (Figure 2) are natural resource management groups which include people's participation in local development decision-making. To support their implementation, the programme had recourse to national consultants. National consultants were chosen by the executive agency among a list that contained three names. Consultants' participatory approach methodology

25 NGOs could hardly be established in the country and for years the EU funds supporting Tunisian associations have been adjourned by the national authorities.

26 The 26.26 was a presidential programme of solidarity that was based on private funds. Each Tunisian had to contribute.

constituted a 'professional identity card' on which they were evaluated and in the case we observed, guaranteed the maintenance of the institutional system of governance (Canesse 2010) characterising the latest Tunisian decades – and the management of the population through local development activities. In such context, the interest in participatory approaches appeared to be closer to the discourse on practices than to practices themselves, and as a consequence ADG illustrated continuity in practices and administrative conceptions. The Ministry of Agriculture and Hydraulics Resources, the Ministry of Interior and Local Development, the Democratic Constitutional Rally and the local notables interacted around the ADG, a local crucial issue.

However, it has to be recognised that memberships can be numerous, diverse, and the actors involved in ADG can sometimes evolve simultaneously in these different circles. In southern Tunisia (and contrary to the governorates closer to Tunis) administrative agents are often native from the governorate and therefore can help at including their native area within development programmes. Widely, memberships can intertwine administration and/or political and/or notability and/or ADG memberships. The limits are vague, and institutional memberships are mobilised according to the situations – or kept quiet.

The Ministry of Agriculture and Hydraulics Resources, administration of execution and 'traditional' mediators of rural development

The Ministry of Agriculture and Hydraulics Resources (MAHR) was established in 1947 and became over the years the main ministry in terms of resources and staff, whose agents are the traditional mediators in rural development, constructing public problem and defining appropriate solutions (Muller 1990). Following Lipsky (1980) the street level bureaucrats belong to the Agricultural Development Regional Commission (ADRC) that is the Ministry of Agriculture's decentralised services in the governorates. However, as the leading players in rural policies implementation, they have a strong knowledge from the ground and for most of them underlined ADG limits during the interviews. Depending on the governorates – and thus actors and context – the MAHR agents had an unequal level of involvement in ADG implementation. The first observed case, in the governorate where the ADRC ranks among those which have the most means, was very instructive. It showed a strong implication of the agents (follow-up and budget correction), as well as a mobilisation of the law as a norm of negotiation and conflict resolution (Lascoumes 1990; Lascoumes and Le Galès 2007). Yet this strict application of the law and of the General Constitutive Meeting[27] proceeding conditions created conflicts with the national consultant whose strategies and methods were based on efficiency imperative.[28] Besides, despite their involvement, the use of the law, the room for manoeuvres appeared to be limited: an ADG

27 The ADG are created during a General Constitutive Meeting.

28 In other words, the consultant's goal is the ADG creation, whatever the proceedings conditions.

was created whereas the Meeting's chairman (head from one of the Ministry of Agriculture's service) refused to sign the constitutive report.[29]

The Ministry of Interior and Local Development, administration of control on territory and population

Whereas it is often justified by economic or technical criteria (e.g. water and soil conservation), the territory is the result of political stakes (Belhedi 1989; Mellor 1989) connected to security, control of the population and administrative actions (Thiebault 1998), and the Ministry of Interior and Local Development (MILD) represents the administration that controls population and territory. With regards to ADG legal norms and implementation, the MILD is the main administration that carries out an *in itinere* control: delegates and *omdas* have a pregnant role, choosing ADG members, dissolving the groups according their activities or their budget.

As it constitutes a representative institution, the territorial stake is at the core of ADG creation. In the governorate of Kairouan, the ADG arouses the interest of a multiplicity of administrative or political actors representing the various territories, and the ADG territory was defined in the delegate's office. In the second case, the ADG reunified two areas and thus enabled the reunification of a sector that had been previously administratively divided. The son of the former *omda* was elected at the head of this ADG. This case also revealed several interventions of the MILD representatives; upstream to the creation, the delegate supported the inclusion of additional zones not to limit the ADG territory to the former sector. He also validated farmers' candidacies to the board committee and chaired the General Constitutive Meeting; the positions into the board committee were distributed in his office. Lastly, once the ADG was created – and the former sector recomposed – the delegate did not deliver the receipt of declaration, preventing ADG of any activities. This non-delivery is another example of the common process at that time for associations in Tunisia (Hibou 2006). In the third case, the *omda* was the most influential actor appearing at each step of the project and in our fieldwork as well.[30] He was highly involved in programmes' activities and in ADG creation, notably because the group represented a possibility to transfer the local water management, which was a source of conflict in these days, to his relatives.

Democratic constitutional rally cells

In each case we observed, the DCR representatives were implicated, notably in the area definition, and played a full part so as to the administrative agents. The DCR cell had a double role in ADG creation. First of all the consultant included

29 ADG must replace former natural resource management groups; here the agent refuses to sign as ADG creation does not lead to the replacement of the existing group but to the superposition of a new local institution.

30 Before official interviews, he first appeared during an 'unexpected' meeting in front of the delegation and without being introduced.

its member as facilitators, in spite of the international methodological guides underlining the necessity not to make political representatives intervene. The DCR could also make available to the development programme a meeting room adjacent to the cell's main one. Whereas the ADG was said 'not to be political', 'to be different' as an animator justified herself (2004), the reality slightly modified her words. Other cases also showed the link between ADG and DCR cells. ADG presidency could be a springboard to reach the DCR cell presidency, so as the cell presidency could also be a springboard for ADG presidency.

The local notables

ADG creation highlighted the monopolisation of local structures by the notables who were sometimes very used to exchanges with national and international institutions of research and development. Indeed, during the interviews the ADG board committee's members obviously mastered international development vocabulary, and clearly identified the advantages they could benefit from the programme. In the first case, the ADG creation lead to the monopolisation of natural resource management by the local notables, a family of big owners whose plots were not included in the programme, and who were at the head of a forest collective management group, which was not dissolved after ADG creation. The conditions of interviews with these farmers were also instructive: they took place in the 'the community house' defined as 'a meeting point for all the natural resource management groups'(Interviews 2006) which was the biggest and the best fitted among the several ADG we visited. It quickly turned out to be a DCR cell, meaning the 'community house' boiled down to the equipment of the DCR cell. This was an additional example of permanency considering the indirect impacts of the Rural Development Program presented above. To some extent programmes enhanced the implementation of the presidential party. The second case reflected farmers' high indifference and the scepticism of the Ministry of Agriculture agents with regards to the population's representativeness through ADG. The head of the group was a big landowner and the son of the former *omda* (before the administrative division); the treasurer, who belonged to the same family, was the DCR cell's president. The ADG meeting point also revealed other local stakes; the land plot on which it was planned to be built also belonged to this family. Moreover the board committee members did not consider organising an annual general meeting, and contrary to the ADG in Kairouan governorate, the president and the treasurer did not intend to recruit a technical director. Therefore the ADG appeared as a mode of representation of a powerful family only and not as a mean to involve the most excluded in local development decision-making. Lastly, in the third case, ADG creation overlapped a conflict between the current *omda* and the former president of the DCR cell who kept controlling the water system. Finally the *omda*'s nephew and his son were elected in the board committee – and, throughout the ADG, became in charge of water management.

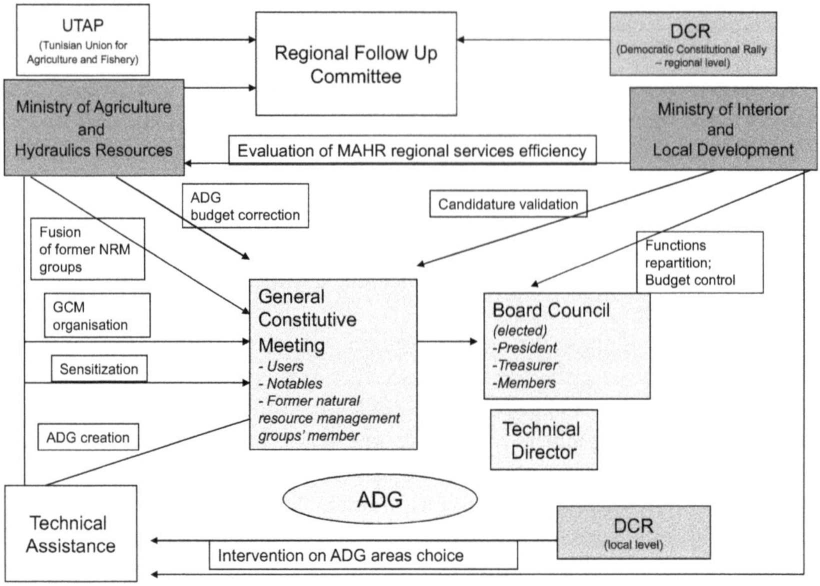

Figure 12.2 Agricultural Development Group's Implementation

Source: Aude-Annabelle Canesse, 2010.

Conclusion

Development associates policy with politics and has to be thought of dialectically. The political context and the institutional environment help at understanding development in the country and the failures of the latter reveal the roots of the Arab Spring. Development policies, tools and plans together involve political and administrative actors and the participation of the weakest remains low. Internal and external factors explain these limits and first of all, the nature of the plans, the procedures, the national context, and the absence of reflexivity and local dynamics analysis in programme-based approach to public policy. Indeed, the programmes constitute a categorisation process of the population which begins during the project formulation through the logical framework approach and which erases the local realities of the field. Moreover they enhance the reproduction of practices that aims at controlling the population and the territory leaving rural areas aside.

As seen above, Tunisian developmental institutions in the rural sector are characterised by the intertwining of the different levels. In addition, the decision-making is located at the central level while local and governorate level councils are only advisory. Nonetheless, administrative institutions are not the only ones: since the Independence in 1956, the regime has been characterised by a single party, the *Neo-Destour* or the Democratic Constitutional Rally, at the crossroad between political party and administration, and representing a key actor at each policy step, from decision-making to implementation.

Large-scale development programmes aiming at improving livelihood conditions in rural areas have been implemented in the rural sector since 1973. The first experience, the Rural Development Programme, encountered limits: it boiled down to thinly spreading, its management was put into question, some of its actions were characterised by inadequacy and broadly were strongly politicised. The following programme, the Integrated Rural Development Programme, illustrated the permanency, in particular with regard to beneficiaries' participation and the politicisation of actions; therefore the situation in rural sector remained worrying. Some change could be noticed in international development programmes, notably in management. It was transferred to non-state actors, this transfer being justified by neutrality and a better efficiency, but this led to an additional process of categorisation due to the several actors (and in consequence procedures) within the programme. Moreover, practices remained similar. Forty years after the first programme, improving livelihood conditions meant fighting against rural exodus and settling populations in underdeveloped areas, in which the users rarely held titles enabling them to access to loans. In addition, the territory had also an international dimension as seen in the governorate of Tozeur, near the border between Tunisia and Algeria. The new oases aimed at both controlling illicit crossings and securing the border delineation. The development programmes implemented by the Ministry of Interior and Local Development and the Development Office of R'Jim Maâtoug, depending on the Ministry of Defence, constituted additional examples of border issues.

Lastly, programmes include politics and local power relationships – and that was what the implementation of Agricultural Development Groups revealed. Participation to the programmes and to ADG highlighted the stranglehold of political and administrative actors and local notables who could combine these three memberships. Indeed the local level is a framework of state's action; as such the very dense territory *maillage* enables the framing and represents an agent of the latter's reproduction, like Maurel (1984) noticed. Local development programmes and ADG forced farmers to get closer to administration, and especially to the Ministry of Interior and Local Development; despite their goals they appeared to be one of the modality of the territorial *maillage*. Local Development programmes and state were essentially two sides of the same coin. While its management had 'officially' evolved, development did not reach its goals: the rural zones remained the poorest, most of their inhabitants remained the most excluded, and the country has remained deeply marked by territorial inequalities, opposing rural and urban areas.

Following 14 January 2011, the Democratic Constitutional Rally was dissolved, the Ministry of Interior was no longer in charge of the 'local development', and the creation of associations was liberalised, all these representing as many strong political signals. After three years, a new Constitution was promulgated in January 2014 that takes into consideration the questions of governance, territorial representativeness and inequalities. The Second Republic of Tunisia: a route for policy and practice change?

References

Andersson, K., 2009. Orchestrating regional development through projects: the 'Innovation paradox' in rural Finland. *Journal of Environmental Policy and Planning*, 11(3), pp. 187–201.

Baduel, P.R., 1985. La production de l'espace national au Maghreb'. In: Baduel, P.R., ed. *Etat, territoire et terroirs au Maghreb*. Paris: Editions du CNRS, pp. 3–47.

Belhedi, A.,1992. *Sociétés, Espace et Développement en Tunisie*. Tunisia: Publications de la Faculté des Sciences Humaines et Sociales, Université Tunis 1, série Géographie, XXVII.

———, A., 1989. Le découpage administratif en Tunisie. *Revue de Géographie du Maroc*, 13(2), pp. 3–25.

———, A., 1992. *L'organisation de l'espace en Tunisie. Production et reproduction?*. Tunis: Publications de la Faculté des Sciences Humaines et Sociales, Université Tunis 1, série Géographie, XXVIII, p. 270.

Canesse, A.-A., 2010. Rural 'participation' and its framework in Tunisia. *Journal of Economic and Social Research*, 12(1), pp. 63–88.

———, 2012. To be or not to be … political? Development discourse, ticipatory instruments and programme-based approach in Tunisia. In: Sjöblom S., Andersson K., Marsden T. and Skerratt, S., eds. *Short-termism and sustainability. Improving governance in spatial policy interventions*. Farnham: Ashgate, pp. 35–55.

———, 2014. *Les politiques de développement en Tunisie. De la participation et de la gouvernance sous l'*ère *Ben Ali*. Paris: Les Editions des Archives contemporaines, Etat du Droit et de la Démocratie.

Craig, D. and Porter, D., 1997. Framing participation: development projects, professionals, and organizations. *Development in* Practice, 7(3), pp. 229–36.

Dumoulin, L., La Branche S., Robert C. and Warin P., 2004. *Le recours aux experts. Raisons et usages politiques*. Grenoble: Presses Universitaires de Grenoble, p. 479.

Giovalucchi, F. and Olivier de Sardan, J.-P., 2009. Planification, gestion et politique dans l'aide au développement: le cadre logique, outil et miroir des développeurs. *Revue Tiers Monde*, 198, pp. 383–406.

Grémion, P., 1976. *Le pouvoir périphérique*, Paris: Seuil, p. 477.

Hénia, A., 2006. *Etre notable au Maghreb. Dynamique des configurations notabiliaires,* Paris: IRMC, Maisonneuve & Larose.

Hibou, B., 2006. *La Force de l'obéissance. Economie politique de la répression en Tunisie*, Paris: La Découverte, p. 362.

Interviews, 2004, 2006. Different persons in charge of development.

Kovách, I. and Kučerova, E., 2009. The social context of project proliferation – The rise of a project class. *Journal of Environmental Policy and Planning*, 11(3), pp. 203–21.

Lascoumes, P., 1990. Normes juridiques et mise en œuvre des politiques publiques. *L'Année sociologique*, 40, pp. 43–71.

Lascoumes, P. and Le Galès, P., 2007. *Sociologie de l'action publique*. Paris: Armand Colin.

Lequesne C., Rivaud P., 2001. Les comités d'experts indépendants: l'expertise au service d'une démocratie supranationale? *Revue française de science politique*, 516(6), pp. 867–80.

March, J.G. and Olsen, J.P., 1989. *Rediscovering institutions*. New York: The Free Press.

Maurel, M.-C., 1984. Pour une géopolitique du territoire. L'étude du maillage politico-administratif. *Hérodote*, 33–34, pp. 131–43.

Mellor, R.E.H., 1989. *Nation, State and Territory. A Political Geography*. London and New York: Routledge.

Perroux, F., 1964. *L'économie au XXe siècle*. Paris: Bordas, p. 321.

Ratzel, F., 1988 [1897]. *Géographie politique*. Paris: Editions Régionales Européennes, p. 385.

République tunisienne, Ministère du Plan, Commissariat général au Développement regional, 1983. *L'évaluation du Programme de développement rural, 1973–1982*.

Rondinelli, D.A., 1976. International assistance policy and development project administration: The impact of imperious rationality. *International Organization*, 30(4), pp. 573–605.

Sethom, H., 1992. *Pouvoir urbain et paysannerie en Tunisie*. Tunisia: Fondation Nationale pour la Recherche Scientifique, Publications du CERES, p. 393

Sjöblom, S., Andersson, K. and Godenhjelm, S., eds, 2006. *Project Proliferation and Governance: The Case of Finland*. Helsinki: Swedish School of Social Science, University of Helsinki, SSKH Meddelanden, 69, p. 100.

Sjöblom, S. and Godenhjelm S., 2009. Project proliferation and governance – Implications for environmental management. *Journal of Environmental Policy and Planning*, 11(3), pp. 169–85.

Thiebault, J.-L., 1998. Les relations entre l'Etat et la périphérie. In: Ben Salah H. and Marcou G., eds. *Décentralisation et démocratie en Tunisie*. Tunisia: L'Harmattan, Logiques Juridiques, pp. 79–98.

Chapter 13
Conclusion: The LEADER Colours on the Democracy Palette

Leo Granberg, Kjell Andersson and Imre Kovách

In the following chapter, we will summarise the findings in terms of the impact on LEADER of the tradition of democracy, or lack of such a tradition, in the countries investigated in this book. LEADER, as described in the introduction of this book, is the central element in European Union's rural development policy since 1990's and an important pioneering experiment of new ways of governance, in general. In this book we have asked whether LEADER has had a positive effect on democracy, especially in countries with an authoritarian tradition. While answering this question, we will also draw conclusions regarding the ultimate question; whether LEADER really improves the situation in rural areas or whether the benefits are rather reaped by those with power? We are aware of the problems in this endeavour and that the answers run the risk of being rather superficial. While the diversity of the individual case studies on EU's rural policy does not allow a rigorous comparison of the results, we find it important to summarise the findings of these studies. Generally, to be able to answer the questions posed above we have to operate with two variables: the context, in terms of political traditions and practices that LEADER is implemented within, and the effects of LEADER work over time against the backdrop of this context.

The categorisation of countries and settings in terms of democracy is an uneasy task; the main pitfall is to adopt a Western centred perspective blindly, seeing democracy in North-Western Europe and Northern America as the model for democracy everywhere. The existing democracy in these countries have its obvious flaws, furthermore comes the debate about the superficial audience democracy (Manin 1997), the theoretical attack from the post-liberal theorists on liberal democracy (Sørensen and Torfing 2005) and e.g. the Occupy movement, all expressing a concern on the state of democracy in the West. Democracy and popular steering in countries and cultures, not normally in focus, is also worth taking into consideration, especially since they may embody dimensions that eschew the dominant discourses. Thus, Switzerland has its special democratic features where referendums can overthrow the aims of the political elites, micro-societies like the Åland Islands between Finland and Sweden (autonomous within the republic of Finland) are fertile ground for grassroots democracy, and even insularity in itself has proved to be a factor furthering democratic development in non-Christian communities (Anckar 2008). There are also interesting types of civic cultures in a wide range of societies, even though they are not usually primarily built upon the autonomous individual. Nevertheless, in today's Europe, there is a wide variation between different countries and societies in terms of democratic conduct, as is confirmed in several of the chapters in this volume. Thus, the account on Russia by Leo Granberg et al. (Chapter 7) is focused on finding signs of functional bottom-up activity in a setting dominated by top-down bureaucratic steering and local passivity. Dénes Kiss and Enikő Veress (Chapter 10) demonstrate the difficult influence of ethnicity on partnership mobilisation in Romania, a country not equipped with a system to equalise the situation, while Bernadett Csurgó and Imre Kovách (Chapter 4) see signs of old and new elites trying to capture LEADER and its resources in Hungary. At the same time, the accounts from Denmark, Finland, UK and Italy presuppose a democratic context where LEADER turns out to be a challenge too, taken often as a deviation from conventional democracy rather than an intervention with a potential to improve it, which, on the contrary, is the perspective in Central and Eastern Europe – and might be so in Tunisia.

The analysis of the democratic impact of LEADER sketched above is far from straightforward. In addition to the different points of departure in democratic terms, the onset of the LEADER activities varies greatly and must likewise be taken into account. In Spain, Italy, Denmark and the UK, LEADER was implemented from the beginning in 1991. In the other countries the activity is coupled to the accession to the European Union, for Finland in 1995, for Hungary in 2004 and for Romania in 2007. However, as revealed by the different chapters, the story of the onset is not that simple, since different sorts of preparatory programmes were implemented before LEADER and parallel programmes have been implemented in some countries in the earlier stages of LEADER, for example POMO, the specific rural development project in Finland, which could have boosted early effects. Nevertheless, the onset of the LEADER activity correlates strongly with the differences in democratic preparedness which is strongly related to contemporary political history in Europe, and especially the enlargement of the European Union. On the one hand, democratic preparedness offers an additional measure for the impact of LEADER since considerable impact would be given much weight in a situation of both unfavourable background factors and late onset. On the other hand, LEADER may also have an initial effect due to its novelty as in Spain (Chapter 3) while it has another 'natural history' in the long term with, for example, increasing bureaucratisation which is indicated in several of the chapters.

Among the early adopters of LEADER, Spain is a noteworthy case. In terms of democratic traditions and preparedness, the country must be seen as a mixed example since it was under Franco's rule until the mid-1970s and thus has quite new democratic institutions. Accordingly, LEADER seems to have played a crucial role, for example in creating a regional institution for inter-municipal cooperation. Also in other respects LEADER seems to have fostered local and regional cooperation and capacity building, for instance through genuine involvement of private actors in the LAGs, According to Javier Esparcia et al. (Chapter 3), a special positive feature in Spain is the role of thematic working groups within LEADER which have widened the scope for cooperation and networking considerably. At the same time, in Spain, as in several Western European countries, a 'backlash' was observed in LEADER during the programme period 2007–2013, because of bureaucratisation and even dismantling of the specific institutional structures built up for LEADER during the earlier periods. All in all, the Spanish account still shows a positive balance despite recent concerns.

The account on the Danish LEADER work by Annette Aagaard Thuesen (Chapter 5), more than any other chapter in this volume, goes into the question of the new sort of local democracy versus the old one, and what has been gained and what has been lost. Thuesen regards it as a progress that the Danish LAGs have adopted principles of representativity on their boards through democratic elections among their members, although within the confines of quota prescriptions (citizens, enterprises, organisations, public authorities). Still, LEADER democracy is the democracy of the active and engaged. In terms of internal work, the Danish LAGs resemble the conception of integrative democracy as portrayed by March and Olsen (1989): there is an emphasis on discussion, consensus-building and learning. The LAGs show therefore, according to Thuesen, a development towards a mix of integrative and aggregative principles, channelling of diverse interests towards optimal compromises and solutions - in a sense taking the best from both worlds.

The importance of the more democratic elections to the LAG boards, which thus adhere to the principles of aggregative democracy, is underscored in the account of women and the LAGs that Thuesen makes together with Petra Derkzen (Chapter 8): the democratisation of the entrance channels to LEADER in Denmark has improved the representation and position of women on the LAG boards. This is not a general principle since the more rural and 'traditional' the LAG is the fewer women enter the board. However, in more modern and urban settings, especially young and well educated women gain entrance to the boards. And this entrance is not confined to the rank and file since the elected women often gain positions as vice chairman, secretary and treasurer, if not so often as chairman. Thuesen and Derkzen suggest, on the basis of their findings, that 'input legitimacy' and open channels are better for women than traditional affirmative actions since they often cement the position of the underprivileged. The risk remains that these measures too can trigger counteraction from those who feel their positions threatened. The principal argument in these two chapters is that the elements of post-liberal democracy – bottom-up, partnerships, networks, direct participation, etc. – in LEADER may not be the crucial element in LEADER's contribution to democracy but rather the combination of these elements with the principles of liberal democracy.

Similar questions are highlighted in the account on LEADER's democratic features in Finland and the UK, explicitly using an elaborated version of March and Olsen's dualistic scheme of aggregative and integrative democracy as a frame of reference (Johan Munck af Rosenschöld and Johanna Löyhkö, Chapter 2). Here, it is noted that the entrance channels to LEADER are restricted in both countries, depending on a range of factors from 'auto recruitment' of the board to passivity and lack of interests among those who could be engaged. However, once inside, the board interaction is quite lively and largely follows the principles of integrative – or deliberative – democracy. A question, which recalls many of the arguments in the liberal versus post-liberal debate on democracy, is therefore whether factors such as a lack of formal barriers, intensity of engagement, mix of knowledge and viewpoints, just-in-time interventions, etc., may compensate for the drawbacks in terms of the ideal models of representative democracy, or if they even add up to something superior to the old model?

Munck af Rosenschöld and Löyhkö question whether the pronounced integrative features of the LEADER work are due to the limited access to it, which can also be interpreted as exclusion, and here they get strong support from Marko Nousiainen (Chapter 6), who analyses political peculiarities of the LEADER work in Finland. According to Nousiainen, the features of consensus and integration, which characterise the internal LAG work in Finland are not natural borne but rather characteristics in LEADER, which are created and forced into the practice by important external actors in cooperation with dominating LAG members. There is simply a storyline that LEADER should work in a certain way and persons who express opposing views on different matters are regarded as 'troublemakers' and are frozen out, to the detriment of any real democracy, according to Nousiainen. This out-with-the troublemaker mechanism is the most severe exclusion mechanism in the present LEADER systems. It seems important to combat this mechanism with the widening and legitimising of the entrance channels to LEADER, for example through democratic elections, as has been done in Denmark.

Giorgio Osti (Chapter 9), for his part, sees a sort of cleavage among Italian LAGs with a considerable part of them functioning according to aggregative principles; municipalities, organisations and 'vested interests' competing for the resources LEADER brings with it and using for example the territory only as a marketing device. On the other hand, there are also examples of more integrative LEADERs and LAGs where the actors work together in order to strengthen local resources and capacity and adhere to the principles of endogenous development. Osti asks for an emphasis on content in the LEADER work, instead of only focussing on participation and distribution. He also questions the dual model of aggregative and integrative democracy and stresses the importance of resources to participate, which can even out the differences in power and resources now *de facto* excluding many legitimate stakeholders.

Among the newer LEADER countries, Hungary is characterised by a somewhat mixed situation. Csurgó and Kovách (Chapter 4) describe the general tendency of both the old and the new local elites to try to capture LEADER and the resources it brings with it. Especially pertinent here is the new 'project class' (Kovách and Kucerová 2006 and 2009). In spite of this, the authors state with High and Nemes (2007), that the EU accession 'has led to a significant growth in the influence of civil society by inducing the state to involve NGOs and civil organisations in decision-making processes'. A feature described in Csurgó's and

Kovách's three case-studies is the strong bureaucratisation of LEADER with the supervising agency MVH in some cases rendering the LAG 'redundant and without a real function'. In one of the cases, the LAG itself seemed to avoid conflicts by distributing LEADER money to as many applicants as possible, thereby missing its own function of prioritising and allocation. Another interesting case showed strong activity by churches, indicating a considerable involvement by society at large in the LEADER work.

Ildikó Asztalos Morell (Chapter 11) investigates the Hungarian LEADER from another angle. She focuses on the attempts within a regional LEADER to enhance the living situation of Roma groups and improve their and the neighbourhoods' environment in general. The means to achieve this was on the one hand a biogas project, where biogas would be produced from biomass gathered by poor Romani. On the other hand, Romani would take care of the general garbage services both in order to get the job done, to enhance their living conditions and to empower themselves through important and respectable societal functions. However, of these two initiatives only the biogas project seems to have gotten air under its wings, unfortunately without much involvement from the Roma communities. The garbage project has been stranded by the attitude from the local authorities that things are functioning well as they are, with professional entrepreneurs gathering and transporting the garbage. Asztalos Morell points out the weak resources of the Roma population, whose interests do not coincide with those of the dominant actors. Sometimes entrepreneurs belonging to the minority can find a win-win situation acting in tandem with a minority collective, but this was not the case in this situation.

The effects of ethnic cleavages are also played out and studied by Kiss and Veress in Chapter 10 concerning the formation of a LAG in the Transylvanian macro-region in Romania. At the beginning the activists from the Hungarian ethnic minority took on the task of forming a LAG in the region in collaboration with an experienced entrepreneur in rural development projects. This is crucial since the Hungarian minority is considered quite active in civic matters, in a society generally characterised by political top-down measures and passivity at the local level. However, the events took a completely new turn with steering 'signals' from central political actors – and money to be gained in forming a LAG – and suddenly an urban-based firm organised a much larger LAG in close cooperation with dominating, national and political interests. This is actually the harshest example of LEADER being captured by interests above the local and regional level and used for reasons quite different to those stated in its 'philosophy'. The two latest mentioned cases open the question whether EU has sufficient methods in use to correct such practises and avoid them in future, or if it do not have?

The two accounts on LEADER-like activities in the non-EU countries, Russia and Tunisia, tell about attempts at bottom-up initiatives in societies and political cultures characterised by top down steering and low level of autonomous activity in local communities. The story from Tunisia (Chapter 11) is much about 'old wine in new bottles': the authorities claim to introduce new types of so-called bottom-up rural development programmes in the country, but in reality during the implementation these are transformed into measures that suit and serve the echelons in political and economic power position. According to Aude Annabelle Canesse, 'the power and beneficiary relations … [are stable] despite changing programmes'. In Russia, the situation looks a little bit more promising, not least because of the active role of (often) well-educated, middle class women in local development issues. According to Granberg et al. (Chapter7) the role of women was also central in the Lake Ladoga region case study, where the local development initiative based on LEADER approach was launched and supported by the EU and Finland. It is also promising that the concrete projects often targeted local culture and platforms for social interaction – and, in extension, capacity building. Similar projects are newly also launched in some other Russian rural regions with Russian funding. Local project activity in Russia partly differs from LEADER approach. It is, in any case, impressive to note how local actors are able to overcome the difficulties caused by barriers of bureaucracy and the hierarchical traditions of the Russian state system.

To sum up this review of the different experiences on LEADER, and similar programmes, Denmark stands out as a promising example of LEADER making a difference in a society from the beginning marked by a well-functioning democracy. The clue to this development seems to be, first and foremost, the combination of liberal and post-liberal, aggregative and integrative, democratic principles. The strong integrative features of LEADER work in Finland and in the UK are interesting. However, this may on one hand indicate considerable exclusion of legitimate stakeholders. On the other hand, we do not know whether this seemingly well-working internal integrative democracy gives good results, in terms of different democratic values. The cases of Italy and Spain give mixed impressions, although there are indications of long term positive results of the LEADER work in post-authoritarian Spain. The Italian account focuses on LEADER in the struggle over power and resources but it also gives a glimpse of very promising LEADER work. The Hungarian experiences are likewise mixed but not without promising traits, 'the project class' can be both a curse and a blessing here. However, ethnic divisions are not easy to cope with as is shown in Hungary as well as in Romania. LEADER has proven to be able to support local initiatives and empowerment but it still needs stronger mechanisms to struggle against marginalisation and exclusion, be it for ethnic or other reasons. Certainly, something beyond LEADER methods is needed to improve the situation in European rural areas. The Tunisian case gives a few clues as to whether LEADER-like methods may work in societies not accustomed to bottom-up development work while the Russian experiment gives some ground for hope. In Russia local activity addressing concrete local problems is increasing. Russia's international conflicts and tightening economy did not (yet) seem to have major negative effect on that tendency at least in the autumn 2014.

As a whole, the crucial achievement of LEADER is to delegate planning to the local level, combined with giving opportunities to local stakeholders to make initiatives and to implement them in a democratic spirit and for mutual benefits. For this kind of activity, LEADER needs an institute called 'project'. Projects seem in principle to fit in the practice in very different European political cultures. After all, the most important contribution projects make to societal steering and democracy is to bridge different nations, sectors, cultures and social groups and thereby also to include as many parties and stakeholders as possible (Andersson et al. 2012). This volume shows that projects can facilitate positive changes, if their weaknesses such as short-termism or exclusion of the 'troublemakers' and the weakest stakeholders are taken into account. One of the best acknowledgments of the LEADER method is that it still is alive, and thriving, while most of its fellow "community initiatives" from the 1990s since long are forgotten. And furthermore, that it even has influenced the urban sphere, giving birth to initiatives such as "city LEADERs". This testifies to its methodological, which to a large extent is about democracy, and social and economic development qualities.

References

Anckar, C., 2008. Size, Islandness, and Democracy: A Global Comparison. *International Political Science Review*, 29(4), pp. 433–59.

Andersson, K., Sjöblom, S., Marsden, T. and Skerratt, S., 2012. Conclusion: Progressing from Governance Challenges to Approaching 'Must Hit' Spatial Policy Targets. In: Sjöblom, S. Andersson, K., Marsden, T. and Skerratt, S., eds. *Sustainability and Short-term Policies: Improving Governance in Spatial Policy Interventions*. Farnham: Ashgate, pp. 301–19.

High, C. and Nemes, G., 2007. Social learning in LEADER: Exogenous, endogenous and hybrid evaluation in rural development. *Sociologia Ruralis*, 47(2), pp. 103–19.

Kovách, I. and Kucerová, E., 2006. The Project Class in Central Europe: The Czech and Hungarian Cases. *Sociologia Ruralis*, 46(1), pp. 3–21.

———, 2009. The Social Context of Project Proliferation – The Rise of a Project Class. *Journal of Environmental Policy & Planning*, 11(3), pp. 203–21.

Manin, B., 1997. *The Principles of Representative Government*. Cambridge: University Press.

March, J. and Olsen, J., 1989. *Rediscovering institutions: the organizational basis of politics*. New York: Free Press.

Sørensen, E. and Torfing, J., 2005. Network Governance and Post-Liberal Democracy. *Administrative Theory & Praxis*, 27(2), pp. 197–237.

Index

References to illustrations are in **bold**

For Product Safety Concerns and Information please contact our EU representative GPSR@taylorandfrancis.com
Taylor & Francis Verlag GmbH, Kaufingerstraße 24, 80331 München, Germany

www.ingramcontent.com/pod-product-compliance
Ingram Content Group UK Ltd.
Pitfield, Milton Keynes, MK11 3LW, UK
UKHW021722240926
475702UK00029BA/1820

* 9 7 8 1 1 3 8 5 4 7 2 8 5 *